D0844230

THE SOW—
IMPROVING HER EFFICIENCY

THE SOW —improving her efficiency

PETER R. ENGLISH
BSc, NDA, PhD

WILLIAM J. SMITH
BVMS, MRCVS

ALASTAIR MacLEAN
SDA, NDA

FARMING PRESS LIMITED
WHARFEDALE ROAD, IPSWICH, SUFFOLK

DEDICATION

To our patient wives
Anne, Lesley and Margaret
and neglected children

First published 1977
Third impression 1979
Second edition 1982

ISBN 0 85236 127 0

Set in ten on eleven point Times and printed in Great Britain on Fineblade Cartridge paper by The Leagrave Press Limited, Luton, for Farming Press Limited.

CONTENTS

ILLUSTRATIONS

PLATES PAGE

DIAGRAMS

PAGE

FOREWORD

by Ken Woolley
Managing Director, Pig Improvement Co. Ltd.

This is a specialist book, another published by Farming Press, and it fills one of the gaps in a comprehensive account of current knowledge of pig-keeping.

There has been an awareness of the difficulties of combining the practical with the academic approach to many farming topics. However, few authors can match the grasp of the scientific and economic aspects of their subject with the practical experience and acute observation shown here by Peter English and his co-authors.

In days of economic pressures, and in an industry accustomed to fluctuating fortunes, the authors are right to remind many of us that we are still not very good at our job, and to indicate the way to make substantial improvements. In some cases this advice involves putting into context several well-known aspects of sow production, but in addition to this the book contains a wealth of information, mainly about husbandry which is not readily available from other sources and should be immensely useful.

This approach is one which has largely been disregarded by academics, perhaps because it does not fall easily into any one scientific discipline. It is therefore particularly good to see one of our centres of higher education and research grasping the nettle.

The authors are not afraid to give firm opinions in fields which allow differing points of view. Greater attention to detail and the welfare of the sow is urged by people deeply involved in their subject. It is advice we should all heed.

Although points are illustrated by UK data, most of the contents are relevant to pig-keeping throughout the world.

The book demands respect of all engaged in the practice, teaching or study of pig production, and if its impact compares with that of the authors' effect in the field on farmers and managers, it must be a success.

K. W. Woolley

Fyfield Wick,
Abingdon,
Oxfordshire.
September, 1977

PREFACE

Both the science and practice of pig production have made considerable strides in the past decade. Knowledge on specific aspects of pig production has increased at an alarming rate and many developments have taken place in practice. It has been extremely difficult for those involved in research and teaching to keep up with the ramifications in the practical situation, while it is equally difficult for the farmer and his staff to keep abreast of developing knowledge, to sift it and incorporate what is most useful and relevant into practical systems.

The aim of this book is to effect an up-to-date remarriage of the science and practice of weaner production so as to provide a basis for, and stimulate, more effective systems of production.

The sow and her output have been subjected to critical examination and the major reasons for shortcomings isolated. Causes of failure are seldom straightforward; they often involve genetic, behavioural, environmental and nutritional factors and the effect of these on the maintenance of good health. The latest knowledge on these disciplines which was considered to be of most practical relevance has been reviewed and intermeshed with our experiences of the great variety of existing commercial practices and systems in formulating the basis for more efficient systems of weaner production.

Pig production has developed and improved over the years by sensible integration of research findings and practical experiences. On this sound basis the industry will continue to develop. At this stage, a handful of producers have effected this integration more efficiently than the general mass and are already achieving very high levels of efficiency. It is hoped that this book will help those who are already very efficient to consolidate and improve even further. However, our most fervent wish is that the great majority of producers who are achieving only 'average' levels of output will find this book a useful source of information and ideas to assist them in improving their output and efficiency and thus place themselves in a much stronger competitive position to meet the future, however severe the challenges may be.

Peter R. English
William J. Smith
Alastair MacLean

Aberdeen
July, 1977

Chapter 1

WEANER PRODUCTION: ECONOMIC ASPECTS

THE AIM of this chapter is to pinpoint the basis of a high level of efficiency in weaner production in both biological and economic terms. The economics is deliberately pitched at a fairly superficial level, designed merely to indicate the link between biological and economic efficiency in weaner production. Those wishing to delve more deeply into pig production economics should consult the many useful publications and textbooks available on the subject.

The sow has one commercial purpose in life which is to produce weaners, and the more efficiently she does this, the higher will be the profit margin on any pig enterprise. However, the general level of profitability in pig production is strongly influenced by the 'pig cycle'.

THE PIG CYCLE

Pigs have a fast rate of reproduction and, in periods of scarcity of pig meat and accompanying high prices, there is a stimulus for existing producers to expand and for former and new producers to go into production. Often there is over-reaction in terms of expansion when pigmeat prices are attractive, with the result that a period of scarcity and high prices is often followed two to four years later by a period of surplus pigmeat which has the effect of depressing prices. So that in a situation where the laws of supply and demand operate without interference from any form of price or output controls, there are fairly rapidly recurring periods of surpluses and scarcities of pigmeat and accompanying low and high prices.

These recurring trends lead to much instability in pig production but the more efficient the producer, the greater the resilience he or she has to periods when prices are low.

PROFIT AS A MEASURE OF EFFICIENCY

Efficiency of a weaner production enterprise can be measured in various ways but the best overall measure of efficiency is profit.

Profit is derived from two components, i.e. output value *minus* input costs.

Output value is dependent on the number, weight and quality of weaned pigs produced, while costs can be divided into two categories:

1. Fixed costs or overheads. These include such items as labour, building depreciation, machinery and stock replacement costs.
2. Variable or direct costs. These include feed, veterinary costs, heating and bedding.

Fixed costs tend to be more or less fixed regardless of the level of output, whereas variable costs increase with output as illustrated in the generalised Figure 1:1.

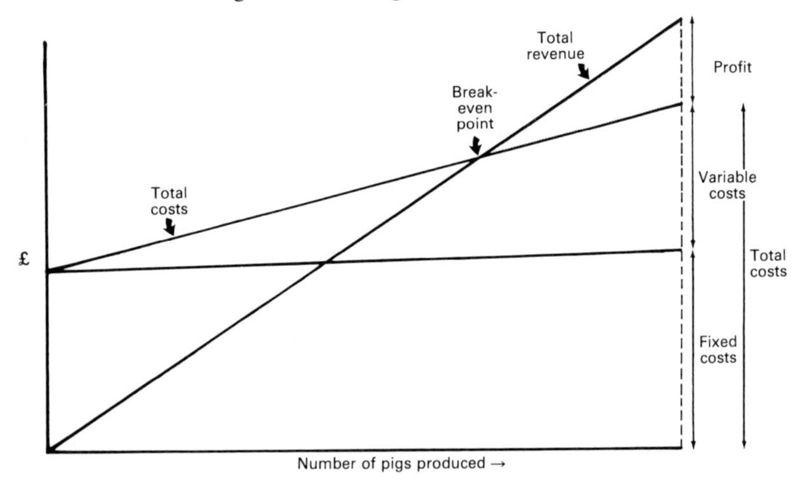

Figure 1·1. Derivation of profit from costs and revenue.

The relationship of the total revenue to the total costs line indicates the level of profitability. The intersect of these two lines is the break-even point, a loss being incurred below this point and an increasing margin of profit above it.

Of the variable costs, by far the greatest is that for feed. Since a large proportion of the feed is required merely to maintain the breeding herd and is therefore independent of output, variable costs are much less influenced by the level of production than revenue; so that, in most circumstances, there is a very substantial incentive to improve output per sow in order to improve profit margins.

The particular level of output required to achieve a break-even point depends on the relative value of pigmeat and input costs prevailing at a particular time. At the worst stage of the pig cycle when surplus pigmeat is available and pigmeat prices are low, a much higher sow output will be required to achieve the break-even point than when the pig cycle is at the other extreme.

So that, at any particular stage of the pig cycle, attempts can be made to improve profit margins by reducing costs and/or by increasing output.

REDUCING COSTS

The factors contributing to total costs in weaner production are as follows:

Factor	*Approximate proportion of total costs (per cent)*
Feed	70
Labour	15
Building overheads, heating, veterinary costs, stock depreciation	15

It can be seen that the greatest scope for cost reductions lies in reducing food costs.

Feed levels for sows have been reduced appreciably from the very high levels prevailing 20–25 years ago and these reductions have been made in most cases without adversely affecting output, i.e. regularity of breeding, numbers weaned per litter and weaning weights.

On some farms, there is probably scope for further reductions in protein and feed levels, the desirable extent of any such reductions depending on the average condition of sows at the time and the extent of their weight gains in successive parities. This aspect is discussed in more detail in Chapter 11. There is scope on many units also to reduce wastage of food caused by such factors as vermin, bad storage conditions and inefficient feeders.

Labour constitutes quite a high proportion of total costs in weaner production, but it is our assessment that further reduction in labour in most weaner-producing enterprises in UK would be a case of being 'penny wise and pound foolish'.

It is also likely that, on some units, the drive to get one man to look after more and more sows could be put into reverse with financial benefit to the producer. This is not to say that we should not explore every possibility of achieving labour-saving systems, but it is likely that on many units it would be very cost effective to intensify labour input at certain crucial stages of the weaner-production process—such as the period immediately after farrowing when it is vital to get the maximum number of piglets well established in order to maximise weaners per litter. Another crucial period is after weaning when efficient heat detection and adequate conditions for, and management of, service is so vital in producing enough litters and weaners per sow per year.

The possibility of reducing overhead costs on housing is rather difficult to evaluate. In planning a new building, the producer is likely to consider several factors simultaneously—the needs of the sow and piglets, capital and depreciation cost per sow, the likely effect on the sow's performance, the reduction of drudgery, provision of reasonable working conditions and reduction of labour requirements per sow.

How much he can reduce costs may depend on the relative emphasis he places on the various factors he considers in planning his building. Probably the greatest scope for reducing building costs per animal lies in making more effective use of available space, while ensuring proper welfare and comfort of stock. In many existing buildings, there is likely to be scope to reduce heating and/or food costs by improving insulation and creating micro-environments for young stock, in particular, rather than being dependent on space heating.

So that it is likely that certain economies can be made on the cost side in most weaner-producing units. However, if the objective is to increase profit margin, it is our contention that there is relatively limited scope for decreasing costs compared to the very substantial scope which exists for increasing output.

INCREASING OUTPUT

On the basis of the sample of pig units participating in the Meat and Livestock Commission (MLC) Pig Plan Recording and Costing Scheme in 1980, the average age at weaning would appear to be about 28 days.

The average output of herds weaning between 26 and 32 days of age in 1980 in this service was as follows:

Number of pigs weaned per litter	9·1
Number of litters per sow per year	2·2
Number of pigs weaned per sow per year	20·1

Source: MLC Commercial Pig Production Yearbook 1980 (April, 1981).

NB: Rounding errors are responsible for weaners per sow per year not being exactly equal to weaners per litter x litters per year.

What is the potential production on a 28 day weaning system? The length of the reproductive cycle will be as follows:

Pregnancy	114 days
Lactation	28 days
Weaning to conception	5 days
Reproductive cycle	147 days

Thus, if all sows, instead of just some, conceived five days after weaning, the number of litters per sow per year would be:

$$\frac{365}{147} = 2.48$$

The average number born alive per litter on 26 to 32 day weaning in herds participating in the MLC Pig Plan Recording and Costing Scheme in 1980 was 10·4. If all these survived to weaning and each sow produced 2·48 litters per year, then the number of weaners per sow per year would be 25·8.

Compared with the existing average of 20·1, this means that there is a shortfall, from what is potentially possible, of 5·7 weaners per sow per year. If, as appears likely, herds participating in the MLC Scheme tend to be the more efficient ones, on average, the difference between the actual national average output per sow and what is theoretically possible is likely to be even greater. What does this considerable shortfall mean in terms of efficiency of food use and profitability?

EFFICIENCY OF FOOD USE

One can calculate, using standard figures, the food requirements of sows on a four-week weaning system at various levels of productivity and use this, along with standard output figures to calculate the efficiency of food use at different levels of productivity. Such a calculation is presented in Table 1·1, while the relationship between sow output and efficiency of feed use is presented in Figure 1·2.

On the basis of the assumptions made, the annual consumption of food per sow averages just over one tonne, the range being from 0·96 tonne for the lowest output sows (14 weaners per year) to 1·06 tonne for those with highest output (26 weaners per year).

It must be pointed out that this food consumption figure per sow makes no allowance for the food consumed as replacement gilts up to service nor does it allow for food consumed by boars or for that consumed by sows between their last weaning and culling. The food requirements for these classes of stock depend on the

TABLE 1·1. Efficiency of food use at different levels of output: four-week weaning system

Weaners per sow per year	*Weaners per litter (a)*	*Litters per year (a)*	*Weight per 4-week weaner (kg) (b)*	*Total weight of weaners at 4 weeks (kg)*	*Total sow food (c) (kg)*	*Total creep food to 4 weeks (d) (kg)*	*Total food (kg)*	*FCR (Total food (kg) per kg weaner)*
26	10·5	2·48	6·5	169	1058	26	1084	6·4
24	10	2·4	6·65	159·6	1041	24	1065	6·7
22	9·5	2·32	6·8	149·6	1025	22	1047	7·00
20	9	2·22	6·95	139	1007	20	1027	7·4
18	8·5	2·12	7·1	128	991	18	1009	7.90
16	8	2·0	7·25	116	974	16	990	8·5
14	7·5	1·87	7·4	104	957	14	971	9·3

ASSUMPTIONS:

(a) That herds which are good in weaners per litter will also be good in litters per year and vice versa.
(b) That piglets from smaller litters will be slightly heavier at weaning than those from bigger litters.
(c) Daily food intake of sows: Pregnancy 2·3 kg
Lactation 2·3 + 0·3 kg per piglet
Weaning to conception 2·3 kg
(d) Creep food intake per piglet = 1 kg.

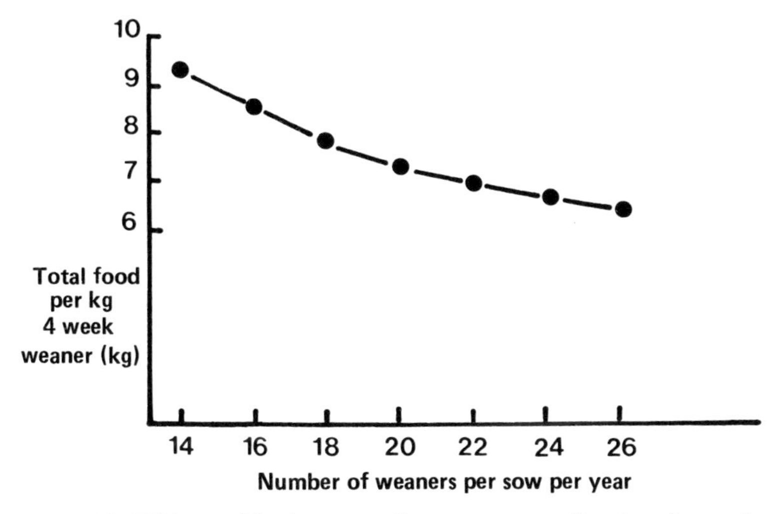

Figure 1·2. Efficiency of food use according to sow output (based on four-week weaning).

efficiency of other aspects of herd management such as age and weight of gilts at first service, promptness of culling and adequacy of boar housing. Estimates of their food requirements have been omitted in the interests of simplicity and also because such omissions do not affect the arguments presented.

To calculate food conversion ratios (FCR) according to different levels of output, sow food and piglet creep food have been added together. Because these feeds do not cost the same per tonne, food cost per kg of gain would be a more useful practical measure of efficiency and this figure is presented later in Table 1·2.

Notwithstanding the differences in price of sow and creep feed, the differences in food conversion ratios according to the different levels of output as presented in Table 1·1 are interesting. As one improves from 18 to 26 weaners per sow per year, FCR improves from 7·9 to 6·4 or by 19 per cent. An improvement from 18 to 22 weaners brings about an improvement of 11·4 per cent in FCR, while even an improvement from 18 to 20 improves FCR by 6·4 per cent.

The need for economies in feed use by avoiding overfeeding of sows is always rightly stressed. However, the considerable indirect influence which improvements in output have on feed efficiency in the sow herd also need to be very strongly emphasised. Sow food is expensive, it is likely to become increasingly so and the most effective way of improving the efficiency of its use is by increasing

TABLE 1·2. The economic significance of improving output per sow on a four-week weaning system

Number weaners per sow per year	*26*	*24*	*22*	*20*	*18*	*16*
Food costs (£)						
Sow (a)	148·1	145·7	143·5	141·0	138·7	136·3
Piglet creep (b)	12·5	11·5	10·5	9·6	8·7	7·7
Weaner diet (4 weeks to 27 kg liveweight)	213·2	195·4	177·8	160·4	143·2	126·4
Total £	373·8	352·6	331·8	311·0	290·6	270·4
Total value of weaners (£) (d)	728	672	616	560	504	448
Food cost per kg weaner (£)	0·53	0·54	0·56	0·58	0·60	0·63
Margin weaners over food (£)	354·2	319·4	284·2	249·0	213·4	177·6
Percentage change in margin over food relative to 16 weaners	+100%	+80%	+60%	+40%	+20%	0
Size of herd required to achieve same total margin over food costs as a 100-sow herd achieving 26 weaners per sow per year	100	111	124	142	166	200

(a) Sow food (14 per cent CP) costed at £140 per tonne.
(b) Creep food (21 per cent CP) costed at £480 per tonne.
(c) Weaner food (19 per cent CP) at £200 per tonne and average FCR of 2:1.
(d) Weaners valued at £1·04 per kg liveweight or £28 per 27 kg weaner.

N.B. 1 Food costs do not include those of boars, of replacement gilts up to conception and of sows between final weaning and culling.
N.B. 2 Prices for food and weaners were the average of those prevailing in UK in May 1982.

sow output. The increases possible in feed efficiency from relatively small increases in output are considerable.

PROFITABILITY IN RELATION TO SOW OUTPUT

The figures presented in Table 1·1 can be converted into economic terms such as margin over food or food cost per unit weight of weaner, by attaching to the physical data, prescribed cost figures for sow and piglet creep food and the value of weaners. This has been done in Table 1·2 using average costs prevailing in UK at the time of going to press (May 1982). To make the data more relevant to the practical situation in which a weaner is often defined as a 27 kg pig of about 12 weeks of age when it is at the point of sale or transfer to finishing quarters, the calculations in Table 1·2 incorporate the costs of taking the 4-week weaner to 27 kg liveweight and its value at this stage.

It can be seen that, on the basis of the assumptions made, improving from 20 to 24 weaners per sow per year reduces feed cost per kg of weaner from £0·58 to £0·54 or by 6·9 per cent; as one increases from 20 to 26 weaners per sow per year, one improves the margin of weaners over food from £249 to £354 or by 42 per cent. Even a relatively small improvement from 20 to 22 weaners per sow per year improves margin over food from £249 to £285 or by 15 per cent.

The crucial importance of annual output per sow is further emphasised if one calculates the herd size required to maintain the same margin over food costs relative to a 100-sow herd producing 26 weaners per sow per year. Relative to such a 100-sow herd, herds producing 24, 22, 20 and 18 weaners per sow per year would have to number 111, 124, 142 and 166 sows respectively to maintain the same margins over food costs on the basis of the data in Table 1·2. Of course, the extra sows required in these less efficient herds would incur considerably higher costs in housing, labour and other miscellaneous costs.

It is accepted that there are economies of scale—larger herds being more efficient in the use of labour and capital than smaller units. However, if herds are already inefficient, attempts to increase, or even maintain, margins should be made by setting out to improve efficiency before considering increases in herd size.

The financial measures of efficiency in Table 1·2 take no account of the other variable costs such as boar food, veterinary charges, bedding, lighting and heating or any of the fixed costs. It is true that, with higher numbers of weaners per litter and more litters per year, there will be slightly greater veterinary costs (e.g. for iron injections) and there may be slightly greater boar overhead costs but, by and large, apart from feed, the costs will not be much increased whether 18 or 26 pigs are weaned per sow per year.

The figures presented in Tables 1·1 and 1·2 are theoretical ones based on the assumptions made. However, their message emphasising the importance of productivity per sow is fully supported by all investigations carried out which have related sow output to measures of efficiency and profitability.

The top 10 per cent of pig herds in terms of pigs reared per sow per year in the Meat and Livestock Commission's Pig Plan Recording and Costing Scheme differ from the rest as indicated in Table 1·3.

TABLE 1·3. Difference between the top 10 per cent of herds and the remainder in the MLC Pig Plan Recording and Costing Scheme 1980–1 (Herds weaning between 19 and 25 days)

Factor	*Advantage for top 10 per cent over rest (per cent)*
Litters per year	+3·5
Born alive per litter	+7·8
Reared per litter	+8·4
Reared per sow per year	+11·8
kg feed per kg of pig weaned	−17·1
Food cost per kg pig weaned	−17·3

Source: MLC Pig Plan Recording and Costing Scheme.

This shows that the top 10 per cent of herds produced almost 12 per cent more weaners per sow per year than the average and this was associated with a 17 per cent reduction in food costs per kg of pig weaned. Data from the MLC Pig Plan Recording and Costing Scheme also forms the basis of Table 1·4.

Table 1·4. Costs and gross margins according to the number of weaners per sow per year (MLC 1980)

	Number weaners per sow per year				
Range:	*Under 17*	*17–18·9*	*19–20·9*	*21–23*	*Over 23*
Mean:	*15·9*	*18·1*	*20·1*	*21·8*	*23·6*
Number of herds	13	31	60	36	10
Weight of pigs produced (kg)	30	31	26	28	29
Feed cost per pig reared (£)	18·44	16·39	14·15	13·78	13·57
Other costs (a) (£)	1·56	2·06	1·79	1·93	1·76
Gross margin per pig (b) (£)	3·11	6·32	7·20	8·52	9·13
Gross margin per sow and gilt (b) (£)	52·42	108·26	156·98	184·02	214·64

(a) Include cost of veterinary services, transport, electricity, water, bedding and other sundry variable cost items.
(b) The gross margins have to cover the fixed cost items of labour, rent, buildings, machinery and equipment.

Source: MLC Commercial Pig Production Yearbook 1980 (April 1981).

Tables 1·2, 1·3, and 1·4 emphasise the importance of output per sow in relation to improving efficiency and increasing profitability.

CONCLUSIONS

It is obvious that the effect of reducing the present shortfall of weaners per sow per year will have an appreciable effect on the profit margin.

Each extra fraction of a pig weaned per sow per year has a large proportional effect on the profit margin because most of the costs in weaner production are fixed regardless of the numbers weaned.

While there certainly is a need to keep costs to a minimum consistent with adequate performance, increasing the number of healthy weaners per sow per year is one of the most worthwhile objectives in pig production. On existing units, it makes possible an expansion of output and margin without expanding facilities and capital investment. On new units, achieving a high number of weaners per sow per year, and the high efficiency of use of feed and other inputs which accompanies such an achievement, is the only feasible way to cover costs, pay for annual instalments on borrowed money and make possible a margin of profit in a difficult economic period.

It is obviously, therefore, in the interests of every pig-keeper to make every effort to improve sow output.

The subsequent chapters examine the many reasons for shortfalls in weaner production and make recommendations on how a pig-keeper can try to eliminate, or at least minimise, these problems and push himself up the weaner league as far as possible in order to achieve extra profits.

As well as making good the deficiencies in existing systems, another consideration is, of course, to reduce age at weaning in order to obtain more litters and more weaners per sow per year. This aspect is covered in Chapter 13.

Chapter 2

THE SOW—ANALYSIS OF HER FAILURES

WE CANNOT pinpoint the exact respects in which the sow fails until we define just what we want of her.

The *credientials of the optimum sow* are as follows:

- Predictable breeding activity and performance as a gilt.
- About 12 strong, normal piglets born.
- 100 per cent survival of these piglets.
- Uniform fast growth to weaning.
- Prompt conception after weaning.
- Predictable weaning to farrowing period.
- Efficient use of food.

In relation to these targets, the average sow fails miserably. Problems that arise are:

1. Gilts often farrow as late as 13–14 months of age and it is usually a case of waiting patiently for the gilt to come in heat rather than controlling the time at first breeding.
2. On average, litter size of gilts is low.
3. Sows can farrow from nil (e.g. if all stillborn) to 20 livebirths in a litter.
4. Birthweights within a litter can vary a great deal.
5. Rearing capacity of the sow can vary from nil (if she dies or goes off her milk) to 16.
6. The suckling positions on the udder are not equally popular or productive.
7. Weaning to conception interval can vary from two to seven days and usually much more.
8. Gestation period can vary from 109 to 118 days.
9. Problems (7) and (8) lead to difficulties in achieving good batch farrowing.
10. Problems (3), (4), (5), (6) and (9) lead to high losses of young piglets.
11. Problems (4) and (6) lead to variation in weaning weight.
12. Often long weaning to conception intervals are experienced.

13. The above inefficiencies lead to inefficient use of expensive food and other inputs.

The average levels of sow performance in UK are likely to be roughly as shown in Table 2·1.

TABLE 2·1 Performance of 'average' UK sow *(See Note 1)*

Age of gilt at first farrowing		= 12 months
Litter size (gilt)	born alive	= 9·5
	reared	= 8·0 (or less)
Litter size (sow and gilt)	born alive	= 10·4
	reared	= 9·0
Percentage mortality of livebirths		= 12·8
Litters per sow per year		= 2·2
Average interval weaning to conception on 29 day weaning	(See Note 2)	= 23 days
Weaners per sow and gilt per year		= 19·8

Note 1 The figures are based mainly on the averages of herds participating in the Meat and Livestock Commission (MLC) Pig Plan Recording and Costing Scheme.

Note 2 On the basis of the sample of farms in the MLC Pig Plan Recording and Costing Scheme, the average age at weaning in UK appears to be 29 days. The interval from weaning to conception of 23 days has been calculated on the basis that the sow will be pregnant for 250 days and lactating for 64 days per year if producing only 2·2 litters per year on 29 day weaning. This leaves 51 days when the sow is not earning her keep and this figure divided by 2·2 litters gives the equivalent of 23 days from weaning to conception. Part of this extemely long interval may be caused by delays in culling following weaning but such sows are still 'passengers' in the herd, using expensive resources for no return.

Thus, unnecessarily long unproductive periods in gilts before they produce their first litter, low litter size born to gilts, high pre-weaning mortality, long unproductive periods in the reproductive cycle leading to a low number of litters per sow per year are all contributory factors to the relatively low output and efficiency of the average UK sow.

However, we cannot blame the sow for all of this failure as she is 'as clay in the hands of the potter'. We are the potters or the pig-keepers under whose influence the sow is and should be able to control her more effectively.

Regarding the shortfalls in sow productivity, we as breeders either are not applying available techniques to the full or else we require new and better techniques to improve sow output and efficiency.

For some of the problems contributing to shortfalls in sow

output, technical solutions may be available but they may not be applied to any great extent because:

(a) They are not considered to be sufficiently practical for easy application.

or (b) The pig-keeper may not have the time to apply them.

or (c) The pig-keeper may not be sufficiently aware:

(1) That a problem actually exists;

or (2) that an existing problem is worth solving;

and/or (3) that the effort necessary in applying an available technique in solving a problem is cost effective or socially acceptable.

For other problems in weaner production, insufficient knowledge may be available on the nature of that problem to allow definite recommendations to be made. However, the fact remains that the stock-keeper is faced with it and must draw on his or her practical experience and the available scientific knowledge in helping to reduce the effect of the problem.

As established in Chapter 1, the financial incentives to improve sow output are very considerable as output is so closely connected with efficiency and financial return.

The crucial areas in improving the efficiency of weaner production are as follows:

The gilt—Achieving greater control over age at first breeding and farrowing and improving numbers born and reared.

Weaners per litter—Achieving greater control over numbers born and reducing piglet deaths.

Litters per year—(a) Shortening weaning to conception interval.

(b) Weaning age considerations.

More efficient use of food.

These are the challenges with which we are faced in order to rectify the shortcomings of the sow and ourselves as pig-keepers.

These vitally important aspects are covered in subsequent chapters.

Basic to an efficient weaner producing enterprise is a good health status in the herd and stock which are sound and efficient from a genetic viewpoint. The important considerations for ensuring a healthy herd and one that is genetically sound and efficient form the basis of Chapters 3 and 4 respectively.

Chapter 3

DISEASES OF THE SOW AND PIGLETS

In this chapter an attempt is made to outline and discuss some of the more important diseases which may affect sows and piglets. Emphasis has been placed on control and preventive measures and also on the circumstances which precipitate or increase the severity of some diseases. The diseases and disorders covered are listed below in the order in which they appear.

1. Genetic disease/Developmental abnormalities.
2. Colibacillosis.
3. *Clostridium perfringens* Type C.
4. Exudative epidermitis.
5. Streptococcal meningitis.
6. Haemophilus infections.
7. Infectious atrophic rhinitis.
8. Navel ill.
9. Joint ill.
10. Teno-synovitis/arthritis due to floor conditions.
11. Haemolytic disease.
12. Purpura.
13. Navel bleeding.
14. Necrosis teats;
Necrosis vulva;
Necrosis knees.
15. Coccidiosis.
16. Transmissible gastro-enteritis.
17. Aujeszky's disease.
18. Vomiting and wasting disease
19. Rotavirus infection.
20. Congenital tremor Type A II.
21. Erysipelas.
22. Epidemic diarrhoea I.
23. Epidemic diarrhoea II.
24. Swine dysentery.
25. Osteomalacia.
26. Foot and leg conditions of the sow.
27. Cystitis/nephritis.

28. External parasites.
29. Internal parasites.
30. Stress.

Note: Farrowing fever and fertility problems are covered in Chapters 9 and 10 respectively.

DISEASE—THE BEGINNINGS

For many years it has been generally assumed that a diseased pig must have succumbed to an infectious agent. As the industry has intensified, our understanding of the term 'disease' has changed completely. Muirhead, 1978, defined 'disease' as follows: 'Disease/disorder may be said to be present in a unit when the normally accepted biological processes fail, with reduction in performance as a sequel'.

It is now recognised that disease may be present in many different forms. It can be more insidious in its onset, more difficult to detect and can be modified in its character by multifactorial components. The ability of the pig to survive the many different insults it will encounter during its life will depend on the balance between a number of forces as indicated below:

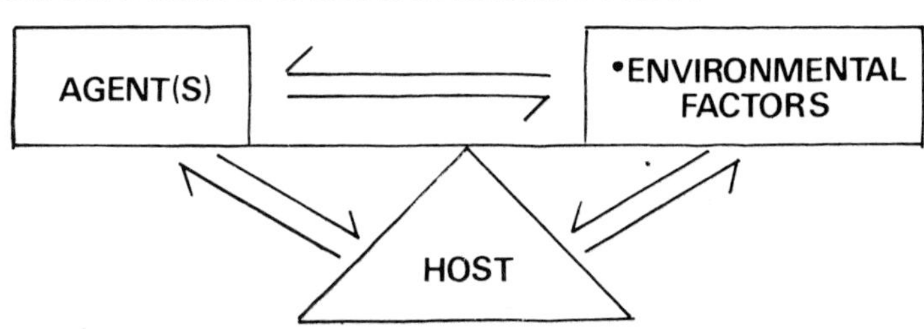

Figure 3·1. The epidemiological triad.

*Environment is used in its widest sense—including such factors as management and nutrition.

Source: J. Walton, 1978—Liverpool Vet School.

The clinically normal pig is simply one which is in a state of balance between the forces within and without. When the defence mechanisms are overcome and the physiological processes fail, the pig is said to be diseased.

Disease-causing organisms may enter the womb so that piglets may be born infected, affected or both. Infectious organisms which gain entry to the womb may cause death of the foetuses, disease of the foetuses or no disease at all. The foetus has a poorly developed defence mechanism and is unable to react to disease until very late in pregnancy.

Although the boar's penis is usually grossly contaminated with organisms, the womb is extremely resistant to infection at the time of heat (hormonal influence). Nevertheless, some organisms do gain entry at service and persist. It has also been established, contrary to previous thinking, that pregnancy to full term is possible despite the presence of organisms such as *E. coli* and Streptococci in the womb, but the greater the number of organisms present, the smaller the number of live foetuses at full term. For further reading on foetal disease, see Chapter 10.

It is therefore important to distinguish between disease contracted at or just after birth and disease contracted within the womb. Disease of the womb and its contents may weaken the foetuses so that they succumb more readily to other infections after birth. Foetuses may also be affected with genetic and developmental abnormalities.

GENETIC DISEASE AND DEVELOPMENTAL ABNORMALITIES

Man has always had a primitive dread of deformity in his own species and has probably extended this fear to embrace similar happenings to his animals. Because of this, he tends to react emotionally rather than rationally to anything that looks abnormal.

What is genetic disease?

Genetic disease may be interpreted as any departure from morphological or functional normality of the individual which is conditioned by the operation of its genetic material The development of all the cells in the body is governed by a blue-print known as the genetic code. Abnormal factors in this genetic code

may be inherited from the parents;

may arise by mutation within the ovaries or testes of the parents;

or they may arise by accident within the fertilised egg itself.

Thus, because of such mutation and accidents, all genetic disease is not necessarily inherited, although once it has occurred it is usually heritable, i.e. passed on from generation to generation. Genetic disease is not always congenital (present at birth), e.g. pityriasis rosea. Congenital malformations are not always genetic in origin, e.g. developmental abnormalities due to vitamin or mineral deficiencies.

Developmental disease

This occurs when environmental factors influence the

development of the embryo through various metabolic pathways and produce defects which are clinically, and sometimes pathologically, indistinguishable from inherited defects. For example, maternal Vitamin A deficiency may produce hydrocephalus which can also be produced by a semi-lethal recessive gene which is inherited. Nutritional deficiencies, drugs, toxins, maternal infections and embryological accidents may all produce developmental abnormalities similar to those seen in some inherited diseases. This is known as phenocopying. See Chapter 10 for further reading.

INHERITED GENETIC DISEASE: CAN IT BE CONTROLLED AND ERADICATED?

Some inherited diseases can be eradicated quite easily but, unfortunately, those inherited disorders which are most common in the pig population are not amenable to eradication. However, their prevalence can be decreased.

The following data is required before action can be taken to deal with suspected inherited disorders:

- A correct diagnosis must be made.
- The mode of inheritance must be established.
- The incidence of the defect within the population must be known.
- The benefit of the sire in economic terms must be assessed in relation to the replacements available.

Source: Dr J. T. Done, MAFF

Diagnosis

In many cases, a correct diagnosis can only be made by a veterinary surgeon after completing certain biochemical, pathological and other examinations.

Mode of inheritance

The exact mode of inheritance is not known for many of the common inherited defects. When a defect is controlled by one gene only or one pair of genes, it is said to be inherited in a simple Mendelian fashion (monogenic situation). However, even this situation may be complicated by the fact that the full effect of some genes is not realised. The full effect may be modified by the degree of penetrance (equals frequency of manifestation) and the degree of expressivity (equals the strength of manifestation). Some conditions are linked to the sex of the animal and hence may only be seen in the male, e.g. congenital tremor type A3. Many conditions are controlled by a number of genes with individual small but additive effects. Thus, the mode of inheritance of many defects is complex (polygenic situation).

Incidence
Good records are essential in order to assess the prevalence of a genetic defect within a population. Epidemiological data relating to age, number affected, breed affected, etc., is useful, if not vital, information, not only for helping with diagnosis but also as a basis for control measures.

Economic considerations
Where AI or tested boars are being used, the value of that sire in terms of progeny performance will be known in relation to boars of average genetic merit. Sometimes, it is more economical in the long run to keep a good boar which is the carrier of a known defect rather than replace it with a genetically inferior boar which is not a carrier of the same defect. Without accurate records the total cost of a particular defect would be unknown.

When does genetic disease show up?
Some genetic defects are lethal while others are extremely mild in nature. When genetic disease appears in the embryo (from fertilisation to day 35) the result is usually death and reabsorption. There is evidence to suggest that genetic defects account for a large proportion of the death of embryos. When affected embryos manage to survive the first month of gestation, they usually survive to full term. They are protected in the womb and can tolerate levels of functional and morphological abnormality which leaves them highly vulnerable after birth, e.g. splayleg. Some genetic diseases do not appear until after birth, e.g. pityriasis rosea and asymmetric hindquarters syndrome.

What can be done?
The presence of genetic disease does *not* mean that steps must be taken to control or eradicate it. The methods of control applied will depend on:

(a) The mode of inheritance of the disease;
(b) The level in the pyramidal breeding structure of the pig industry in which they are to be applied, i.e., steps taken at nucleus level may be different from those taken at commercial production level. The factor governing any decision at the end of the day is economics.

CONTROL
(Commercial and some multiplier herds only)

Monogenic defects
Points to note:

- Only one or two abnormal progeny are necessary to establish that a boar is a 'carrier' of the defect.
- 'Dominant' defects (assuming complete penetrance and expressivity) would appear in the male before he was used.
- Sex-linked conditions are not carried by the males. In other words, the females are the carriers and would be likely to give birth to defective offspring irrespective of the boar used.
- Recessive genes are more difficult to deal with because of the carrier status in both sexes. When a carrier boar has been detected, the decision to remove him or not would depend upon the incidence, seriousness, pathogenicity and the availability of other sires of proven ability and known status. The higher the incidence of a defect, the bigger the response will be to the removal of carrier boars.

Polygenic defects
For practical purposes, all conditions which have been unequivocally demonstrated not to have a simple Mendelian inheritance at a single place are included in this category. The mode of inheritance is complex and therefore not amenable to eradication.

For control purposes, all these defects which are not monogenic with complete penetrance and expressivity may be tackled in the same basic way but with variations for different levels in the pig breeding pyramid.

For many of these defects there is a threshold at which the animal will succumb. Both internal and external circumstances influence this threshold value. Splayleg is a good example of such a disorder. The severity of this disease will depend upon such factors as birthweight, muscular development, the extent of myofibrillar hypoplasia, the muscular activity required to establish social order, the litter size and the slipperiness of the floor.

Once an inherited disorder has been diagnosed it must be costed. Well-kept records are, of course, vital for this procedure. The value of the boar (if he has been tested) relative to a boar of average breeding merit, will also be known. Should the disease be costing more than the gain in pig performance attributable to the boar in question, then he should be replaced by a boar free of the unwanted trait and of equal genetic merit. It is very difficult to obtain a boar completely free of known defects. This approach to genetic disease control is called the *cost/benefit approach* in this book. If in doubt a veterinary surgeon or husbandry officer should be consulted.

The more common inherited disorders are described below and the control methods indicated.

SKELETAL DEFECTS

Thick leg

This is a massive thickening of the forelegs.

Incidence: Unlikely to rise above 0·1 per cent—mainly in offspring of the Landrace breed.

Mortality: 5–10 per cent are likely to die or require culling.

Mode of inheritance: Polygenic.

Control: Incidence low—no action.
Incidence high—cost/benefit approach.

There are many other inherited skeletal defects but the incidence is low and they are of no significance in economic terms.

SKIN DEFECTS

Pityriasis rosea

This disorder becomes apparent at eight to twelve weeks of age and disappears four to six weeks later. The disease begins as red, pea-sized, slightly thickened spots which gradually increase in diameter leaving an area surrounded by a raised bluish/red periphery. The lesions resemble ringworm in the human to some extent.

Incidence: Rarely rises above 0·3 per cent. Both offspring of Large White and Landrace affected.

Mortality: Nil.

Mode of inheritance: Polygenic.

Control: No action need be taken. Weaners have usually recovered by the time they are offered for sale.

Dermatosis Vegetans

This condition is also known as club foot. It appears only in the Landrace breed and affected piglets usually survive until five months of age when they develop a fatal pneumonia.

Incidence: Low.

Mortality: 100 per cent.

Mode of inheritance: Simple Mendelian (recessive).

Control: Cull sow and boar.

NERVOUS SYSTEM

Congenital Tremor Type A III (Myoclonia congenita)
This disorder is present in the piglets at birth and is characterised by rhythmic tremors of the head and limbs which fade when the animal rests and pass off when it sleeps. In severely affected pigs, the legs vibrate on the floor so rapidly that they seem to be dancing.

Incidence: Up to 0·1 per cent.
Mortality: 100 per cent.
Mode of inheritance: Simple Mendelian. A recessive/sex linked gene. Carried by dam only. Female piglets are normal, while 50 per cent of male piglets are affected.
Control: Cull dams of affected piglets.

Congenital Tremor Type A IV
Symptoms similar to those of type A III. This disease has only been diagnosed in British Saddlebacks.

Incidence: Up to 0·1 per cent.
Mortality: 100 per cent.
Mode of inheritance: Simple Mendelian. A recessive gene.
Control: Cull boar. Replace with one proven not to be a carrier.

Meningocoele
This condition is present at birth and is characterised by an opening in the skull through which the membranes of the brain protrude. A percentage of affected piglets die in utero.

Incidence: Up to 0·1 per cent. Mainly in offspring of Landrace.
Mortality: 100 per cent.
Mode of inheritance: Polygenic.
Control: Apply cost/benefit technique.

MUSCLE DEFECTS

Splayleg

This is probably the most important inherited disorder of piglets in economic terms, particularly in the Landrace breed. The name of the condition is quite descriptive. The hindlegs are usually more severely affected and, consequently, the affected piglets find difficulty in competing for a teat. If piglets achieve a regular suckle and survive the first four days of life, the condition improves and gradually disappears. Several factors influence the severity (see introduction on genetic disease, page 31). One report has suggested that splayleg may be caused by feeding sows with meal containing toxins from a fusarium mould. To date, these findings have not been confirmed by other workers. It has also been suggested that choline deficiency in the dam may be responsible for the birth of splaylegged piglets.

Incidence: Up to 1·5 per cent.
Mortality: Approximately 50 per cent if left untreated.
Mode of inheritance: Polygenic.
Treatment: Legs may be held closer by tape or elastic. Artificial feeding is beneficial. The majority of piglets alive three days after birth will survive.
Control: Apply cost/benefit technique.
Increase birthweight of piglets (see Chapter 11).

There are other muscular defects such as the acute stress syndrome (metabolic myopathy) which affects finishing pigs and the asymmetric hindquarters syndrome which is of low incidence.

ALIMENTARY (GUT) DEFECTS

Atresia Ani (inperforate anus or 'blind gut')

This condition is self descriptive. Piglets may live up to three weeks of age, but unless treatment is thought worth while, affected piglets should be destroyed at birth. Treatment is easier when the rectum extends right up to the skin. In the female, the rectum sometimes extends into the vagina, forming a 'cloaca' and, in most cases, is usually closer to the skin, making surgical treatment easier.

Incidence: Up to 0·35 per cent.
Mortality: Males 100 per cent.
Females 50 per cent.
Mode of inheritance: Polygenic
Treatment: Surgical.
Control: Apply cost/benefit technique.

REPRODUCTIVE DEFECTS

Inguinal hernia (scrotal hernia)

This defect mainly affects the male but does little to inconvenience the pig except for a few cases of acute or partial strangulation of part of the alimentary tract. Most of the deaths are attributable to accidents at, or subsequent to, castration.

Incidence; Up to 0·7 per cent.
Mortality: Males 0·6 per cent.
Females less than 0·5 per cent.
Mode of inheritance: Polygenic.
Treatment: Surgical (doubtful if economically justified).
Control: Apply cost/benefit technique.

Inherited liability to disease
There is some evidence to suggest that offspring of certain boars may be more liable to succumb to infectious disease. At least it has been demonstrated conclusively that the offspring of some boars are more susceptible to certain serotypes of *E. coli* which were pathogenic to neonatal piglets. However, it is unlikely that such traits are common and the selection of boars for resistance against disease is likely to remain a 'pipe dream'.

COLIBACILLOSIS
(Baby piglet diarrhoea or Coli scour)

The bacterium, *Escherichia coli,* is an organism found in the gut of all pigs. It is part of the normal flora within the gut along with many other organisms. There are many different types of *E. coli* and those that have been identified have been placed in groups known as serotypes. Serotypes which are known to be pathogenic may be found in the gut of normal piglets. Studies have shown that the normal new born pig may experience waves of infection with *E. coli,* i.e. the gut becomes colonised with various serotypes of *E. coli* in turn. These may be picked up from the sow or farrowing pen.

The coliform population in the gut of the sow multiplies rapidly around farrowing and large numbers are excreted in the faeces. When certain circumstances prevail, some serotypes multiply rapidly in the piglet's gut, release toxins (poisons) which in turn cause cessation of stomach movements and flooding of the lower bowel with fluid from the blood stream. The affected piglets scour, become rapidly dehydrated and toxaemic and will usually die unless treated. From what has been said, it will be obvious that *infection* is not synonymous with *disease.* Some of the factors which influence the severity of the disease are outlined in Figure 3·2.

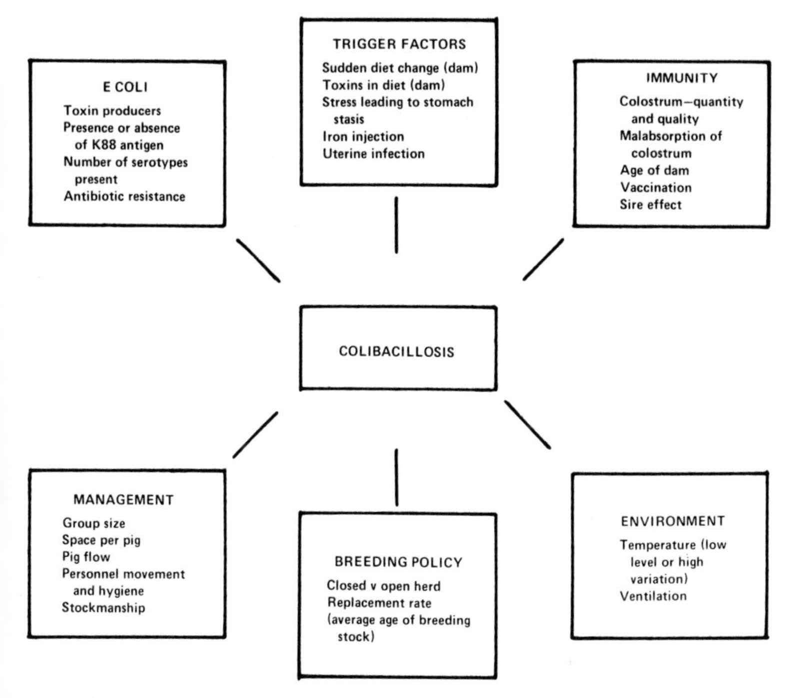

Figure 3·2. Factors affecting severity of disease.

Before discussing these factors in turn several, general points should be noted.

1. The disease accounts for less than 5 per cent of pre-weaning mortality (national basis) but it tends to affect whole litters at a time—hence it is more obvious and worrying to the pig owner.
2. Litters of gilts are more often and more seriously affected.
3. The condition is conspicuous by its absence in properly managed Roadnight (outdoor system) herds.
4. It is more common in large herds buying pigs from many sources.
5. It is more common in large farrowing houses with a continuous throughput.
6. On the whole, it is easier to prevent than to cure.

Control and prevention

The disease may be controlled and eventually prevented by taking steps to:

- Reduce the weight of infection.
- Increase the immunity of the piglet.
- Avoid factors likely to increase the severity of the disease.

Reducing the challenge by infection

Attempts should be made to reduce the intake of new serotypes of *E. coli* to a minimum. Unless the breeding policy dictates otherwise, the herd should be closed except for the occasional introduction of a boar. Only essential personnel should be allowed access to the unit. A changing room for unit personnel should be provided and clean boots and overalls should be available for all visitors including the veterinarian. A good pig-flow is essential. Where possible, farrowing houses should hold no more than 12–15 sows. An all-in-all-out policy with rest and disinfection will lower the prevalence of the disease. All these measures will reduce the challenge by infection.

Increasing the immunity

The immunity of the piglet will depend on the quantity and quality of colostrum received. Once the piglet has started to suckle, the selective absorption of immune proteins from the colostrum rapidly starts to decrease. Consumption of any protein will initiate the shut-down mechanism, hence it is important to ensure that the first dose of protein is not bacteria from the sow dung. Keep the farrowing pen clean! The quality of the colostral immunity will depend on the type and number of organisms to which the sow has been exposed and the length of time for which she has been exposed. Sows, and especially gilts, should be exposed to the bacterial flora of the unit for at least a month before farrowing and for a month before service. A strawed yard or house with a common dunging channel is adequate for this procedure.

This naturally gained immunity may be improved by vaccination of the dam with one of the commercially available *E. coli* vaccines. Sometimes, the serotype involved in the outbreak may not be contained in a commercial vaccine. Your vet may be able to make an 'autogenous' vaccine, i.e., one made from the organism implicated in the outbreak. Sows may also be vaccinated orally by feeding pure cultures of the organism in the drinking water, or food. Whatever method is used, the results expected are not often achieved until a large proportion of the herd has been immunised. Your vet should be consulted about the various methods of vaccination.

It has also been reported that certain boars may sire piglets which are susceptible to the disease. When more than one boar is concerned, your records should provide conclusive evidence if this inherited factor is suspected. Cold, draughty farrowing pens may predispose to the disease. Pens should be mucked out twice daily if the floors are solid (for the first three days). Disinfection between batches may be necessary and is recommended when an outbreak is encountered. Physical and social stress should be avoided as this may lead to stomach stasis, which always occurs before a piglet starts scouring. Do not give iron by injection to piglets which are suffering from colibacillosis (see Chapter 9).

Summary of main points

1. Close herd if possible or buy from one source only.
2. Cut down entry of new serotypes.
3. Adopt 'all-in-all-out' policy with rest and disinfection between batches.
4. Improve hygiene and environment.
5. Increase herd immunity, e.g. pre-service exposure to *E. coli*, commercial vaccine, autogenous vaccine, oral vaccine.
6. Avoid stress.
7. Make sure the diagnosis is correct—other organisms cause diarrhoea. Consult your veterinary surgeon when in doubt.

CLOSTRIDIUM PERFRINGENS TYPE C

This disease usually affects piglets of one to seven days of age and may be prevalent in out-door systems. The causal organism may be found in the soil as well as in the gut of normal pigs and belongs to a large group of organisms which can multiply in the absence of oxygen. The disease is characterised by high morbidity, high mortality, blood stained diarrhoea in the acute form or persistent diarrhoea with wasting in the chronic form. The disease can only be diagnosed by a veterinary surgeon.

Treatment

Antitoxin and antibiotics are recommended, but response is poor once disease has become established.

Prevention

Administer lamb dysentery vaccine to the sows twice before farrowing. (There is cross prevention between *Clostridium perfringens* type B which causes lamb dysentery and *Clostridium perfringens* type C.) In some countries a *Clostridium perfringens* type C toxoid is available.

EXUDATIVE EPIDERMITIS

(Marmite disease; Greasy pig disease)

This condition is characterised initially by reddening and weeping of the skin which gradually becomes thicker looking and blacker as dirt coagulates in the fluid weeping from the lesions. The disease may take an acute, sub-acute or chronic form. As the disease progresses, the piglets may become more listless, hairy and black looking due to the encrustations, resembling peeling sunburn. Eventually, a crust or layer with fissures covers the whole body of the pig. Pigs aged from five to thirty-five days may be affected. There is no evidence of pain or pruritus (itchiness).

Cause

A specific type of Staphylococcus (bacterium) which is commonly found in and around pig skin is said to be the cause of this disease. The organism is harmless unless the skin is broken. Predisposing factors are fighting, castration, ear-tattooing, viral infection or any factor leading to breaking of the skin. Mortality and morbidity vary from 5 to 95 per cent.

Treatment

Antibiotics and cortisone may be administered, and supportive therapy with vitamins (especially B vitamins) is beneficial. This condition may readily be confused with other skin conditions. A veterinary surgeon should be consulted.

Control

Eliminate predisposing causes; practise an 'all-in-all-out' policy; clip teeth at birth; disinfect ear-punching instruments; use sterile knife between litters while castrating and take measures to avoid fighting at weaning by mixing in a large strawed pen with subdued lighting or using the sedative Suicalm (Crown Chemical Co. Ltd., Lamberhurst, Kent). The disease is particularly prevalent during hot, muggy weather in August and September, and management should assure adequate ventilation in these conditions.

STREPTOCOCCAL MENINGITIS

Streptococci are bacterial organisms which may be found in both normal and diseased pigs. There is a wide range of disease conditions associated with these organisms—abscesses, pneumonia, navel ill, arthritis, heart valve disease, dermatitis, metritis and vaginitis to name but a few. In the young pig, meningitis is probably the most significant disease caused by streptococci. Outbreaks have become more common over the last few years, not only in suckling pigs but also in weaners and fatteners.

Clinical signs

Incoordination, peculiar swaying of hind quarters when standing or walking, circling, head held to one side, loss of balance, prostration with paddling movements of the limbs, rotating movements of the eyes and convulsions may be seen in affected pigs. A pig may appear normal in the morning and be found prostrate in the afternoon.

Predisposing factors

The mode of infection is not always easy to determine. Breaks and laceration of the skin, surgical procedures such as castration, infected needles and increased susceptibility with improved lines of stock have all been suggested as factors leading to the introduction of the infection. One particularly virulent type (*Streptococcus suis type II*) can be isolated from the tonsils and pharynx of normal pigs. It may reach the brain via the eustachian tube and the inner ear. Over-crowding, poor ventilation and decreased air volume in flat deck houses due to build up of slurry levels have all been suggested as precipitating factors.

Treatment

The condition must be diagnosed in the early stages if treatment is to be successful. High doses of rapidly absorbed penicillin is the treatment of choice. A sedative should be administered to those with jerky movements or convulsions. Affected piglets must be moved to separate quarters in warm, draught-free conditions. Recovery may take 2–6 days and it is important that they should be given water or glucose saline by mouth or stomach tube until such time as they are able to eat or drink themselves.

Prevention

Until more knowledge of the pathogenesis of the disease is available, preventive steps are difficult to prescribe. However, care should be taken to avoid infection at castration, ear-tattooing, tail docking and other minor surgical manipulations. Teeth should be cut soon after birth to prevent wounds from fighting and any pig badly mauled or lacerated in a fight should be treated immediately with penicillin. Weaned and multisuckled groups of pigs should be inspected closely twice daily for this condition. Although initial reports suggested that autogenous vaccination might be useful, subsequent work has shown vaccination to be relatively ineffective.

HAEMOPHILUS INFECTIONS

Haemophilus bacteria, of which there are several types causing distinct clinical entities, are sometimes found in normal healthy pigs. However, when susceptible pigs acquire infection, mortality may be high. Suckling pigs may develop Glassers disease (*H. parasuis*). The disease is sudden in onset, and usually whole litters are affected at one time. The temperature of affected piglets rises rapidly to 40–42°C (104–107°F), and suckling ceases. Piglets are noticeably lame and the joints become puffy and swollen. Coughing and occasionally nervous symptoms may occur. Mortality may be high and survivors usually become 'runty'.

Outbreaks of Glassers disease may be expected when piglets are born to dams which have been introduced to a herd where the organism is endemic. Affected piglets should be treated with suitable antibiotics (e.g. modern Penicillins) by injection for at least three days. The disease in suckling piglets can be prevented by vaccination of the dam, and to this end a simple formalised autogenous vaccine is effective.

Haemophilus pleuropneumonia is now a well known cause of pneumonia in weaned and fattening pigs, but recently it has been shown to cause a similar type of pneumonia in suckling pigs in the North-East of Scotland. Vaccination of the dam has proved an effective method of control.

INFECTIOUS ATROPHIC RHINITIS

This respiratory disease is one of the most serious diseases ravaging the pig herd at the present moment. The mild form of the disease in suckling pigs (snuffles) is more prevalent than most farmers realise. Many owners only become worried when the disease has reached the atrophic state (loss of structure and function of nasal bones, see Plate 1). A recent survey has shown that atrophic rhinitis is 3–5 times as common now in the UK as it was 10 years ago and it now probably costs the industry in the region of £2½ million due to poorer growth rate and poorer feed conversion efficiency. Source: Prof R. H. C. Penny, London Vet School.

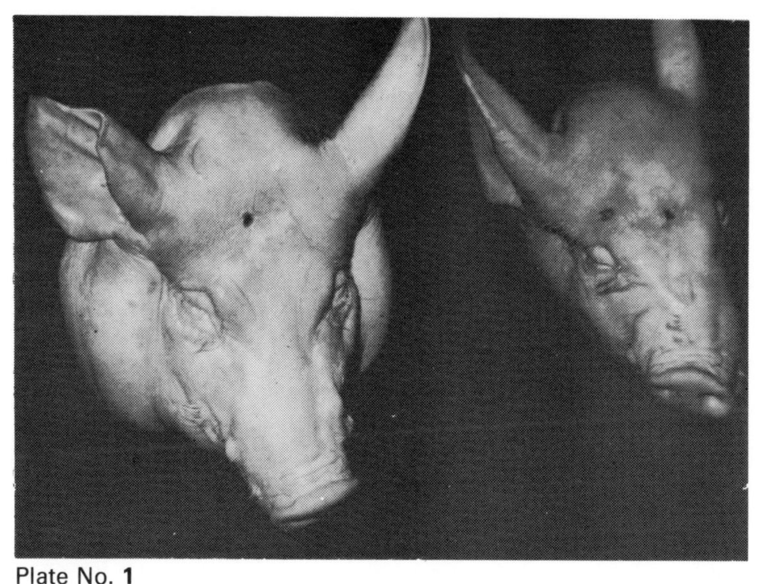

Plate No. **1**

Head from abattoir exemplifying deviation of snout and wrinkling of nose which are typical signs of the atrophic stage of rhinitis.

What do we know about rhinitis?

1. It is an infectious disease of swine primarily spread between pigs of less than 10 weeks of age.
2. There may be one or more infectious agents involved and these are spread via large moisture droplets in the air.
3. It is more common in young herds, i.e., containing many gilts (expanding herds and new herds).
4. It is more common in open herds as opposed to closed herds and in those herds purchasing replacement stock from many sources.
5. It is more common in intensively managed herds where large numbers of young pigs are kept in close confinement often in the same air space.
6. The increase in incidence over the last few years is probably linked with intensification, and improved pigs (this does not mean that all improved pigs are more susceptible to rhinitis).
7. Pigs from some strains of Large White are more susceptible to the disease than Landrace (Landrace pigs can also take the disease).

8. Minimal disease and non-minimal disease herds may be affected but the incidence of rhinitis in minimal herds is lower.
9. Clinically normal sows, gilts and boars may harbour the infectious agents, i.e., they may be carriers of the organisms.
10. The disease is more prevalent during the winter months.

Why should rhinitis be a problem in one unit and not in another even though the same infectious agents may be present in both units? This is a difficult question to answer but nevertheless an attempt is made to explain why this can happen in the following paragraphs.

Figure 3·3 depicts factors affecting the severity of the disease. These factors may differ from unit to unit and hence steps taken to deal with the disease will also vary from unit to unit.

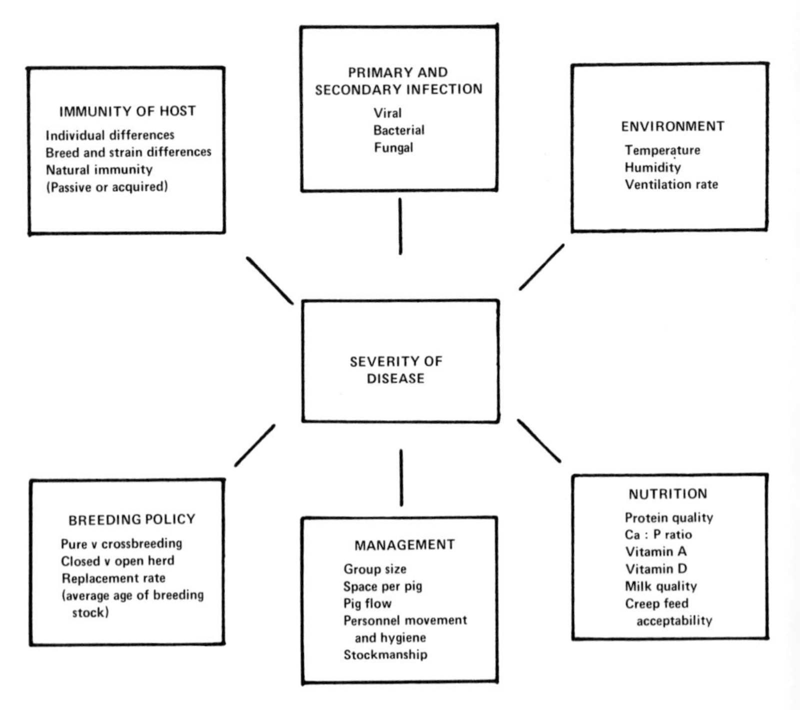

Figure 3·3. Factors affecting the severity of atrophic rhinitis.

The infectious agents
Two organisms are primarily responsible for causing the initial breach in the delicate lining of the nose and upper respiratory tract. One is a bacterium (susceptible to some antibiotics) called *Bordetella bronchiseptica* and the other is a virus (not susceptible to antibiotics) called inclusion body rhinitis virus. The latter organism can cross the placental barrier and, should it do so, newly-born pigs will harbour the infection. These organisms cause more severe lesions when present together or in conjunction with other organisms such as *Pasteurellae* or some of the common respiratory viruses.

Piglets pick up the infection from carrier dams soon after birth and quickly spread it to adjacent litters by droplet infection. Some strains of *Bordetella* release toxins which cause calcium to be reabsorbed from the bones. The turbinate bones within the nose (see Plate 2) are remodelled frequently as this part of the pig's head is growing eight times more quickly than the rest of the skull. Thus, a mild infection with a toxin-produced strain of Bordetella may prevent proper bone formation in the nose.

The virus affects certain cells of the delicate lining covering the turbinate bones. This lining also contains cells which produce mucus and cells with hairs called cilia which all waft in the same direction. Incoming air becomes turbulent within the turbinate bones and is warmed at the same time. Some infectious organisms and dust particles are trapped within the nasal cavity and carried out of the nose in the sticky layer of mucus moved along by the cilia. The nose acts as a defence barrier protecting the lungs from some of these infectious organisms. Infection with inclusion body rhinitus virus will weaken this barrier considerably.

Bordetella organisms tend to cling to the ends of the cilia where they multiply and release toxins. This particular location of Bordetella organisms lessens the efficiency of medical treatment as it is difficult to get enough antibiotic to the organism. Inhalation of the drug would be the choice of treatment for Bordetella but this is impossible for several reasons. Finally, the presence of these organisms within the nasal cavity is not synonymous with disease.

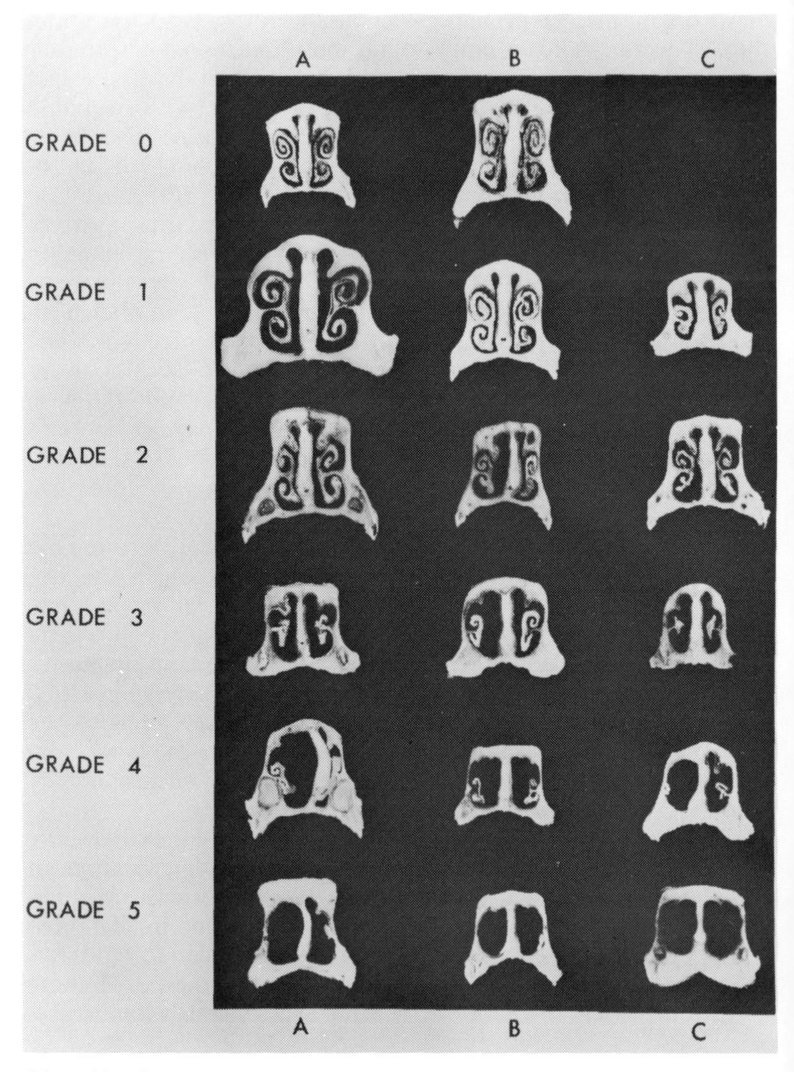

Plate No. **2**

Cross sections of snouts showing normal snouts (Grade 0) and various grades of infectious atrophic rhinitis.

Photograph: Crown Copyright.

Climatic environment

Temperature. Rhinitis is always more severe during winter months. Large variations in diurnal temperature increase the severity of rhinitis. Piglets which have to huddle on top of each other are obviously too cold.

Ventilation. The rate of air exchange plays an important role in the severity of rhinitis. Sneezed-out infective particles should be carried to the exterior rather than being rebreathed by other pigs. The ventilation rate should be as high as can be tolerated (without causing draughts or heat loss from the building to such an extent that heating costs will be prohibitive). Well-insulated buildings make good ventilation less costly.

Relative humidity. Wide ranges in humidity seem to be well tolerated and this facet of environment probably plays little or no role in rhinitis.

Nutrition

Piglets whose noses are blocked with pus and mucus (see Plate 3) cannot suckle and breathe through the mouth at the same time. The intake of milk will be reduced, making affected piglets more dependent on creep feed. However, the sense of *taste* and the sense of *smell* will also be markedly reduced. These factors alone will make affected piglets less willing to consume creep feed. Creep should be *fresh* and of high *quality*. Creep feeders should be emptied once daily and replenished with fresh creep. The creep used should also contain adequate calcium. ARC recommended standards are adequate.

Vitamin A and D intake will also be reduced because of poor milk intake. Both vitamins may be administered simultaneously by injection to suckling piglets. Vitamin A is important for maintaining the integrity of nasal mucus membranes and Vitamin D plays a vital role in the uptake of calcium by the bones. Consult your veterinarian about the use of these vitamins.

Breeding policy

Should gilts be bought from several sources the chances of rhinitis becoming a problem will be higher. The weight of infection will be increased, while the overall herd immunity will be lowered. Buy replacement gilts from one source only and, if necessary, ask for a veterinary certificate indicating the status of health of the herd, especially with regard to respiratory disease. (No one can guarantee animals free of organisms which may cause the disease.)

Several workers have reported that some strains of Large White are more susceptible to the disease than others. The heritability of the disease has been estimated at 0·2; i.e., 20 per cent of the variation in severity is genetic in origin. Landrace pigs are not immune to rhinitis but the incidence of disease in purebred Landrace is low.

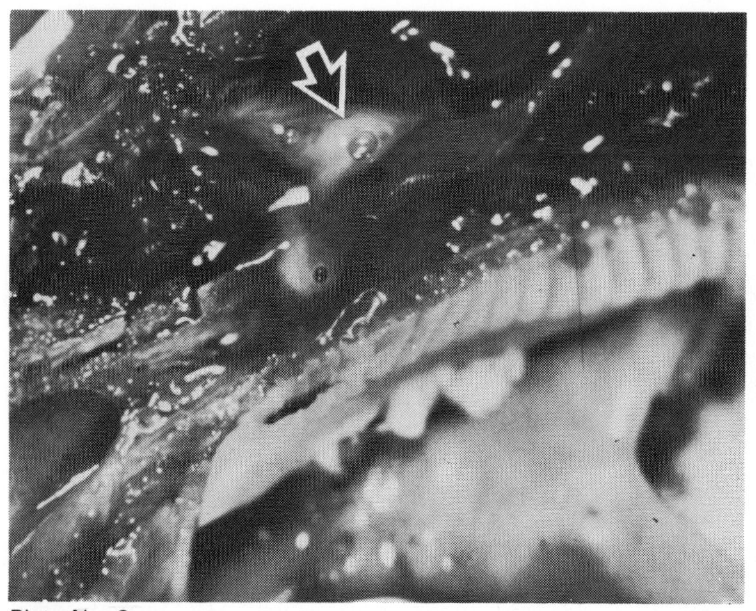

Plate No. **3**

Longitudinal section of nose of 10-day-old piglet showing mucus blocking the air passages between nose and windpipe.

Herd management policy

Units practising continuous farrowing in large farrowing houses within the same air space are likely to have a higher incidence of rhinitis. The prevalence of respiratory disease will reduce when farrowing in small units of 12 to 15 sows, especially when practising an 'all-in-all-out' policy. Where possible, gilts should be farrowed separately from sows. Rest and disinfection of farrowing units will also be beneficial. Attempts should be made to rid the unit of vermin and cats, both of which may harbour Bordetella organisms.

Treatment

The aim of treatment is to reduce the population of bacteria in the upper respiratory tract. As the infection is picked up soon after birth (see Plate 4), this is best achieved by a strategic input of drugs by injection. Piglets cannot be medicated adequately via water or food as they are not consuming enough of either until 4 to 6 weeks of age. The choice of drugs is best left to your veterinary surgeon. Tylosin or potentiated sulphonamides have proved most successful. If necessary, the creep feed should be medicated and feeding of medicated creep continued until at least one week after weaning. The timing, frequency and choice of drug input will vary from herd to herd. In severe outbreaks, 2 to 3 injections are necessary within the first 14 days of age.

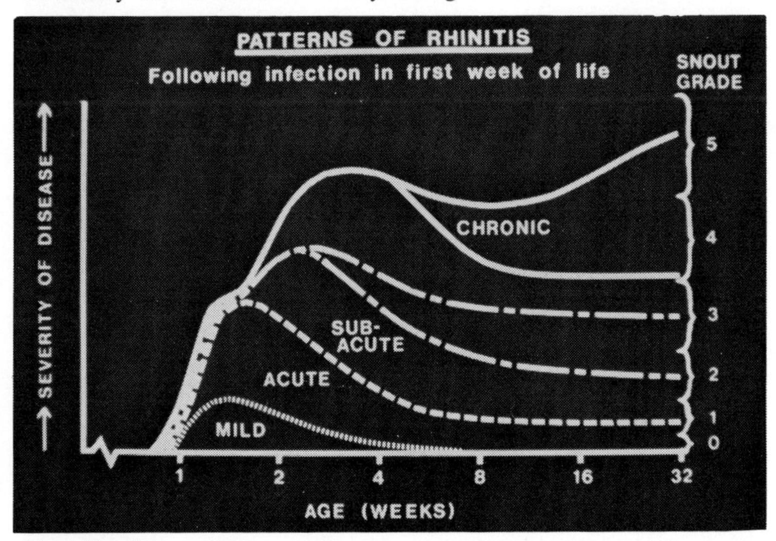

Plate No. **4**

Patterns of rhinitis following infection in the first week of life.
Photograph: Crown Copyright.

Generally speaking, efforts should be made to reduce the weight of infection and increase the herd immunity, while at the same time avoiding factors which increase the severity of the disease. If these measures fail, consideration should be given to total herd replacement with stock from a minimal disease source. Bordetella vaccines are now available and are being used in practice. Reports from the field have been conflicting regarding their usefulness, but the consensus of opinion would suggest that beneficial results may follow vaccination of the breeding herd. However, no guarantee could be given about the usefulness of piglet vaccination because of interference from maternal antibodies.

NAVEL ILL
(Omphalophlebitis)

This disease, as the name implies, is infection of the navel cord or stump. Various organisms may be involved and the incidence of the disease varies greatly from unit to unit. Once infection has become established in the cord, the organism(s) multiply and usually produce an abscess which may be internal or external or both. Toxins produced by the organisms cause the signs of retarded growth, poor appetite, shivering and lassitude. Occasionally, the organisms themselves may invade the whole body via the blood stream leading to septicaemia and death. While in the womb, the blood circulation of the piglet is connected to the placenta (afterbirth) via the navel cord which contains an artery taking the blood out and a vein taking the blood back. The vein is connected to the circulation of the piglet at the liver. It is not surprising that organisms which gain entry to the cord can rapidly make their way to the liver and finally to other organs in the body.

Treatment
Antibiotic injections for 3 to 4 days. Consult your veterinary surgeon.

Prevention
Reduce the challenge by thorough disinfection of farrowing pens, ensuring that the floor has had time to dry before the next farrowing. In small units or very large units where individual attention is available at farrowing, the newly-born piglets should be placed in a box or restricted area under a lamp so that the cord dries up rapidly. Where individual attention cannot be given at farrowing, the cords of newly-born piglets should be sprayed or dipped in an antiseptic solution such as phenolated iodine as soon as possible after birth.

JOINT ILL
(Septic arthritis)

This disease, as the name implies, is infection of the joint cavities (it may be confused with tenosynovitis which is due to infection of the synovial sheath surrounding the tendons which pass over the joints). The disease is often found in association with navel ill and the incidence varies from unit to unit. Many organisms which are commonly found in the farrowing pen may cause the disease and they probably gain entry to the joint from the blood stream of the piglet. The organisms survive particulary well in joints because of the poor blood supply (the blood carries antibodies and cells which destroy bacteria). Organisms may enter the blood stream of the piglet from skin wounds, the gut or the navel cord.

Symptoms
Initially stiffness, disinclination to move, fever, and loss of appetite. As the disease progresses, the joints become enlarged, hot and painful. One or more joints may be affected.

Treatment
High levels of antibiotic for 4 to 5 days. A veterinary surgeon should be consulted. The sooner the disease is diagnosed, the more effective treatment will be.

Prevention
Ensure teeth are cut at farrowing, disinfect pens between litters and treat navel cords. Vaccinate sows against erysipelas.

TENOSYNOVITIS/ARTHRITIS DUE TO FLOOR CONDITIONS

The above problem is now being encountered more often because of:

- Badly laid concrete or concrete laid with unsuitable ingredients.
- The ever-increasing use of slotted/perforated metal floors with sharp void edges and where the void area is too large in relation to the size of the piglets' feet.

Concrete floors
Once the surface has started to disintegrate the aggregate becomes exposed. In many cases the aggregate is composed of sharp-pointed pebbles. The soft tender feet of the newborn piglet are easily lacerated and bruised. This injury occurs when the piglet is actively competing for a teat. Once the sole of the foot has been eroded, infection gains entry and causes septic arthritis of the pedal joints or tenosynovitis higher up the leg.

Perforated metal floors
Many of these floors have sharp void edges and excessively large

void areas in relation to the size of the piglets' feet. Flattened expanded metal with a void size of 43 mm by 14 mm (see Plate 5) is a typical example of such a floor. Injuries occur in three main areas:

- The sole of the foot (see Plate 6).
- Above the heels (see Plate 7).
- On the accessory or supernumerary digits (see Plate 8).

These areas may become bruised and eventually eroded. Once the skin has been broken, infection gains entry to the tissues and causes inflammation and chronic disfigurement (see Plates 9 and 10). Studies in one large herd have demonstrated that the incidence of injury to the hind feet was twice that of the fore feet and that, on average, 12 per cent of injured parts became infected and swollen. The inside claws were injured more often than the outside claws and this is probably a sequel to the downwards, backwards and outwards thrusting movements of the hind legs while the piglets are actively competing at the udder. Partly slotted pens are common but this does not eliminate the problem as most partially slotted areas are at least 1·1 m in length (i.e. parallel to the long axis of the sow). Observations have shown that at least three-quarters of the udder lies on the slotted area and therefore the majority of the piglets suckle on the slotted surface.

Plate No. **5**

Note the sharp V-shaped corners and the large size of the perforations.

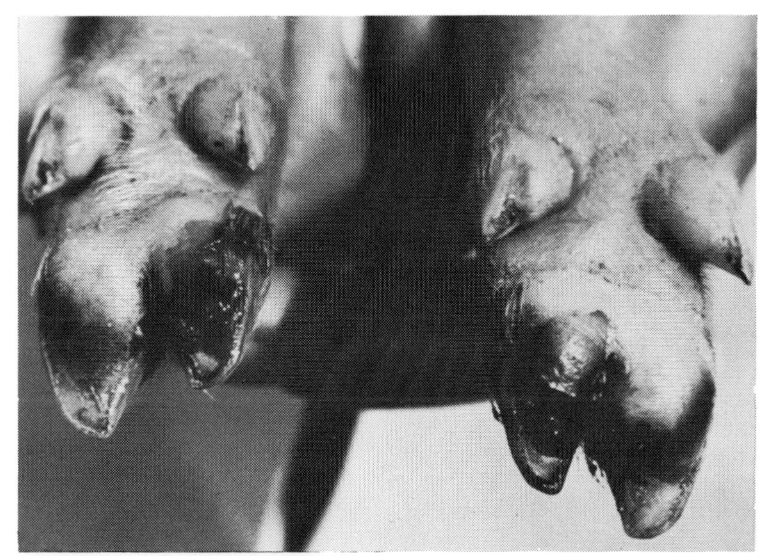

Plate No. **6** Note the erosion of the sole of the claws.

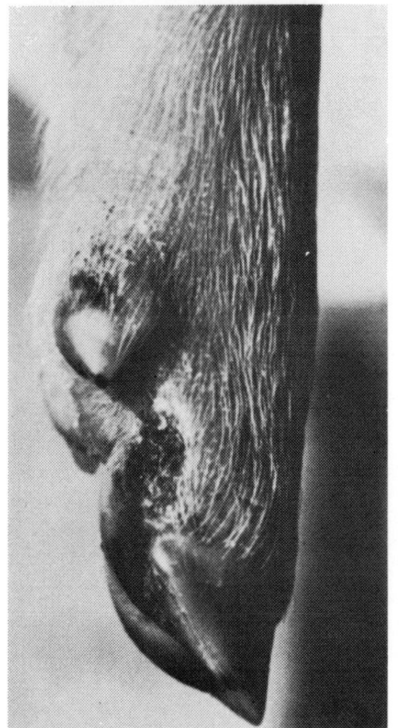

Plate No. **7**

Note the bruising above the heels.

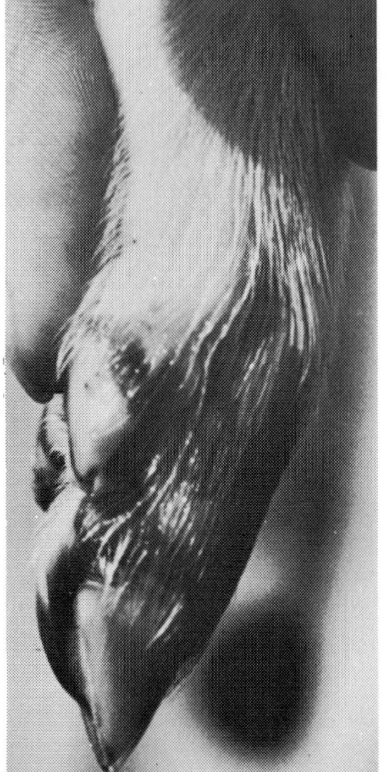

Plate No. **8**

Note the bruising on the accessory digit.

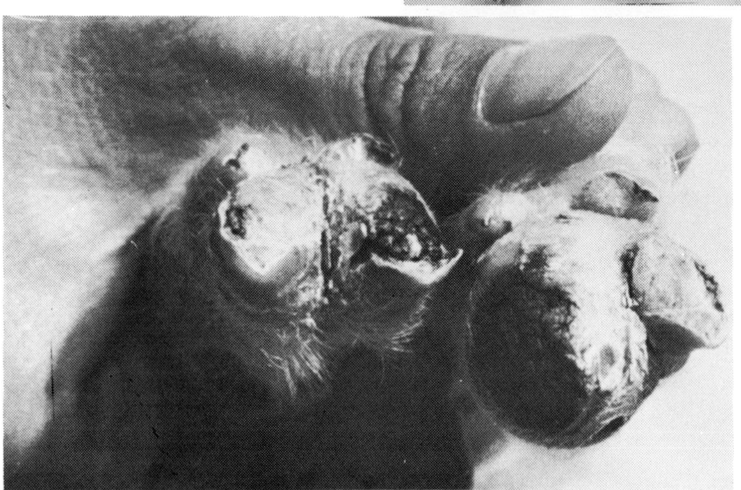

Plate No. **9**

Severely damaged claws.

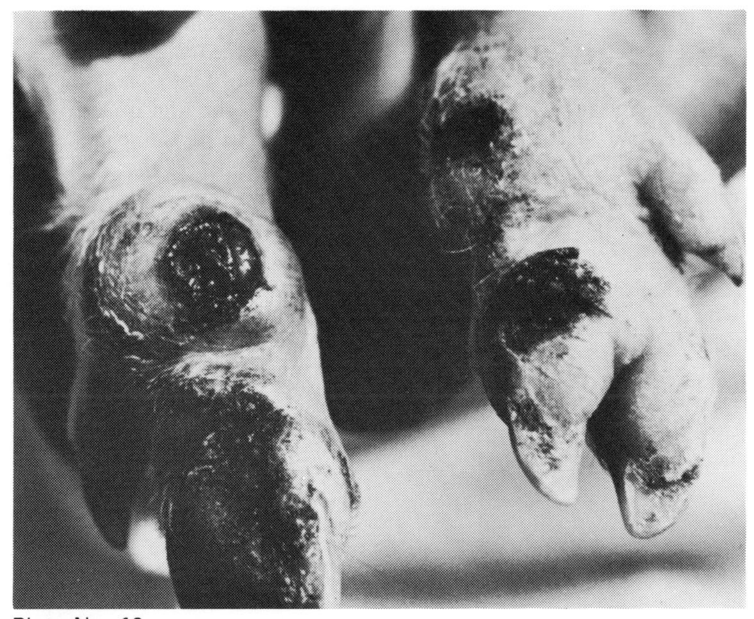

Plate No. **10**

Severely damaged accessory digits.

Prevention

Concrete floors. Either re-lay or paint the surface with an epoxy resin or chlorinated rubber paint. At least a week should be allowed for epoxy resin to dry properly.

Slatted/perforated metal floors. Existing floors may be dipped in two coats of plastic—this reduces the area of voidage and rounds the edges. If this treatment is not available, the floor should be replaced immediately. Replacement perforated floors should have a rounded void edge with the smallest gap no more than 10 mm.

HAEMOLYTIC DISEASE

This is a disorder of piglets of one to five days of age characterised by haemolysis (destruction of the red blood cells), weakness, pallor, occasionally jaundice and death. It is not a common disease. Usually, only one litter at a time is affected and mortality varies from 50 to 100 per cent. The disorder occurs because certain sows become sensitised to the red blood cells of foetuses sired by a particular boar. There is little or no effect on the first litter to this boar. However, should the sow be remated to the same boar she will become hyperimmunised to the red cells of the foetuses. While in utero (in the womb), the piglets are protected because the antibodies from the sow cannot cross the placental barrier. However, the antibodies are absorbed by the piglet once it suckles colostrum and the circulating red blood cells are immediately attacked. The disease can only be diagnosed by a veterinary surgeon.

Treatment
Good nursing and supportive therapy with fluids and vitamins. Keep warm and draught-free.

Action
Cull the sow from the herd. If the sow is very valuable, remate her to a boar which is unrelated to the carrier boar.

THROMBOCYTOPENIA PURPURA

This disorder is more common than haemolytic disease but is still relatively uncommon (0·95 per cent incidence). Nevertheless, it can be a cause of concern because of its nature. It is characterised by pallor, weakness, small purple skin haemorrhages (especially behind the ears and under the belly) and death. Again, only one litter at a time is usually affected and the mortality may vary from 50 to 100 per cent. Piglets usually show signs of the disease at either 2 to 4 days of age or 17 to 20 days of age. The reason for the double peak is not known but at least it has been ascertained that it coincides with a sharp drop in the numbers of circulating thrombocytes (cells vital for blood clotting).

Aetiology
The cause of the disease is hyperimmunisation of the sow against the circulating thrombocytes of offspring sired by a particular boar. As in haemolytic disease, the second or subsequent litters to the same boar are always affected. Piglets die from a multitude of tiny haemorrhages or from secondary infection as they often have a current neutropenia (drop in white blood cells which protect against infection).

Treatment
Keep warm and draught-free, apply fluid replacement therapy, administer multivitamins, trace minerals and antibiotics for three or four days to prevent secondary infection.

Action
As for haemolytic disease. *Note:* If piglets are fostered before consumption of colostrum, the occurrence of haemolytic disease or thrombocytopenia purpura will be prevented.

NAVEL BLEEDING

This disease is characterised by bleeding from the navel cord, pallor and death within 24 hours of birth. If left without attention, the majority of piglets die. The incidence may vary from one or two piglets per litter to a high percentage in a large number of litters. Piglets look normal at birth but rapidly become pale and anaemic within a few hours. The actually haemorrhage from the navel cord is not readily seen, especially on straw or perforated floors.

The disease has not been proved to be due to hereditary factors or the lack of clotting factors in the blood. It may, in fact, be linked to a failure of the cord to constrict after birth. In utero (the womb) the cord connects the piglets' blood circulation to the circulation of the placenta (afterbirth) and blood circulates freely from one to the other. In normal piglets this circulation is shut off (by a process not yet fully understood) immediately after birth at the junction of the navel cord and the abdomen. In navel bleeders this shut-down mechanism does not seem to take place and blood is pumped freely out of the artery in the cord. The disease often disappears from a unit without any action being taken to control it.

Action
Monitor farrowings carefully and tie off any bleeding cords with strong thread dipped in antiseptic. This may mean that the pig-keeper will have to spend many sleepless nights during an outbreak. However, farrowing during the daylight hours may be induced with the use of prostaglandins. Although the disorder may remain at a chronic low level for several years in a herd, sudden outbreaks with high morbidity rapidly diminish and often disappear altogether.

TEAT NECROSIS

As the name suggests, this condition is characterised by death of the tissue forming the teat (see Plate 11). The condition has become more common as herds have become bigger and management more intensive. It has not been seen in Roadnight (outdoor system) herds and this itself would suggest that the condition is linked with hard floors. The condition has been seen in piglets on both solid concrete and metal slotted floors. Necrotic lesions of the tail, hocks, elbows, vulva and under the chin are often found concurrently.

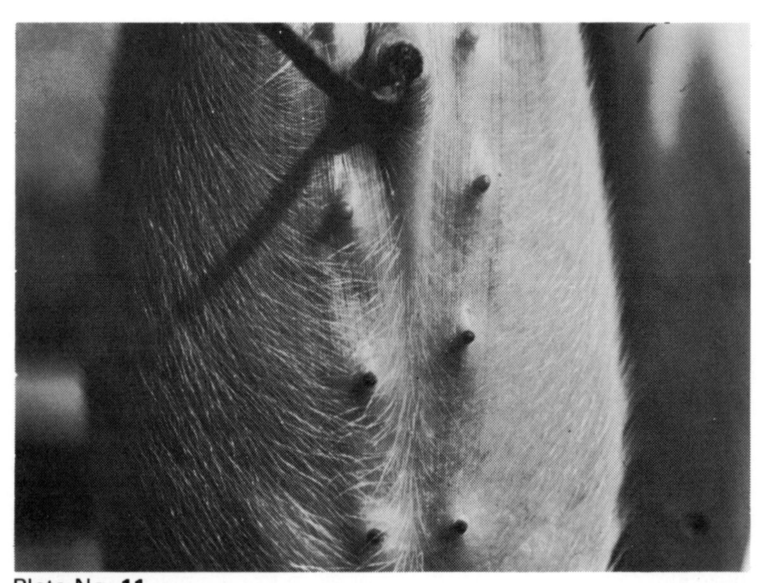

Plate No. **11**

Note necrosis of first two pairs of teats.

Incidence

There is a big variation between herds and the incidence within a herd may be as high as 70 per cent; the ratio of males to females affected is usually one to two. The anterior (front) teats are most commonly involved; teats are usually involved in pairs and up to five pairs may be injured, counting from the front. The higher the incidence of teat necrosis in a herd the greater the number of pairs of teats involved and vice versa. Most of the damage occurs while suckling.

Factors affecting the severity of the condition

- Swelling of the teats at birth is usually present in those piglets which subsequently suffer from teat necrosis. The swelling is probably linked to oestrogen (hormone) levels in the sow at farrowing (see Plate 12).
- New floors with high pH (high alkalinity), rough floors, metal slotted floors with sharp edges to the void, and the absence of bedding precipitate the disease.
- When competition for teats is high, e.g., big litters, poor milk production and poor teat exposure (dam), the disease is more prevalent.
- Intercurrent diseases such as splayleg and congenital tremor (all types) increase the severity and incidence of the condition.

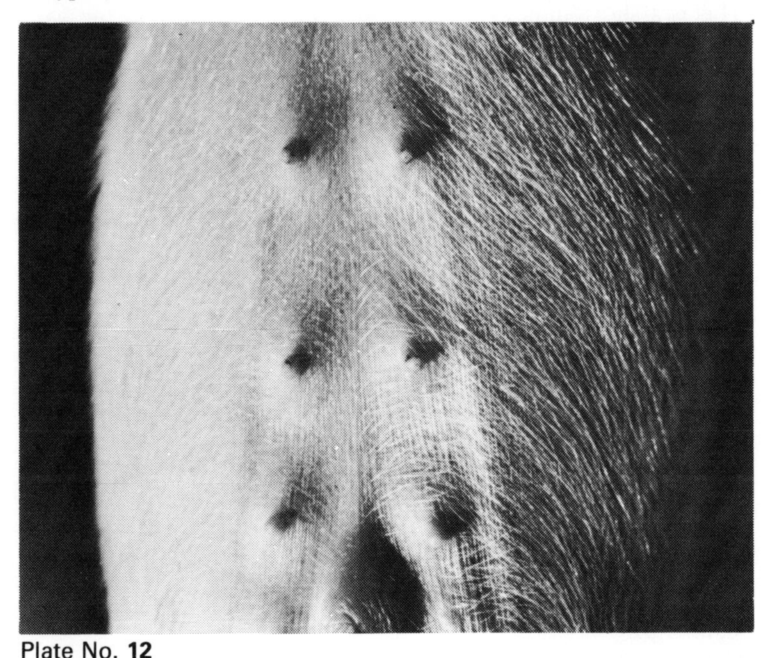

Plate No. **12**

Note swelling of teats.

Treatment
None; once it has appeared, the damage is irreversible.

Prevention
1. Avoid precipitating factor(s) where possible.

2. Apply Bostic or similar preparation to teats of new-born piglets until such time as the causal factor has been identified. *Note:* This condition may be of considerable economic importance in those herds producing gilts as the damaged teats are either absent or non-functional.

INVERTED TEATS

In this condition the teats are completely invaginated like the fingers of a glove turned inside out. The lactiferous ducts are absent and there is no treatment. The condition may be inherited.

PSEUDO/INVERTED TEATS

This is a shortening of the teat due to a constriction at the base. Such teats become normal when suckling begins.

VULVA NECROSIS

This condition is often present when there is a concurrent high incidence of teat necrosis. Swelling of the vulva at birth is always

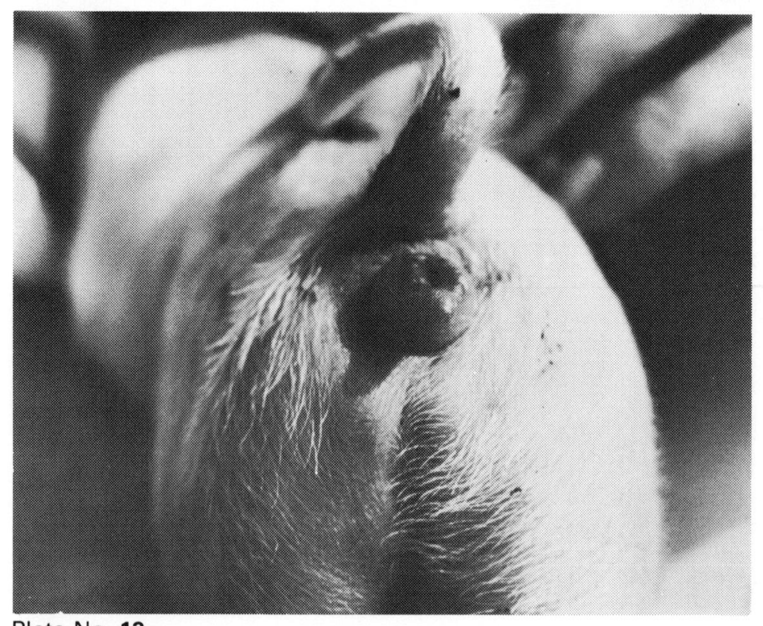

Plate No. **13**

Note swollen vulva.

present in those piglets suffering from this condition (see Plate 13). The vulva becomes bruised, haemorrhagic and lacerated (see Plate 14). As the piglet grows, the lesion heals and constricts. The end result is a misshapen vulva with tiny orifice or no vulva at all (see Plate 15). In a recent survey of a large herd, 15 per cent of female piglets were found to be affected. The factors affecting the severity and incidence of the disease are similar to those for teat necrosis. This condition may be of considerable economic importance in those herds breeding gilts.

KNEE NECROSIS

This condition is extremely common but is of little or no economic importance. It rarely leads to arthritis or tenosynovitis unless the deeper tissues are involved. Any type of hard floor will cause this condition; the severity of the lesions can be reduced by the provision of bedding, but it must be very deep bedding. The cost of supplying the bedding (sawdust is not suitable as it is rapidly scraped away) and removing the dung is likely to be far greater than any drop in performance due to the lesions.

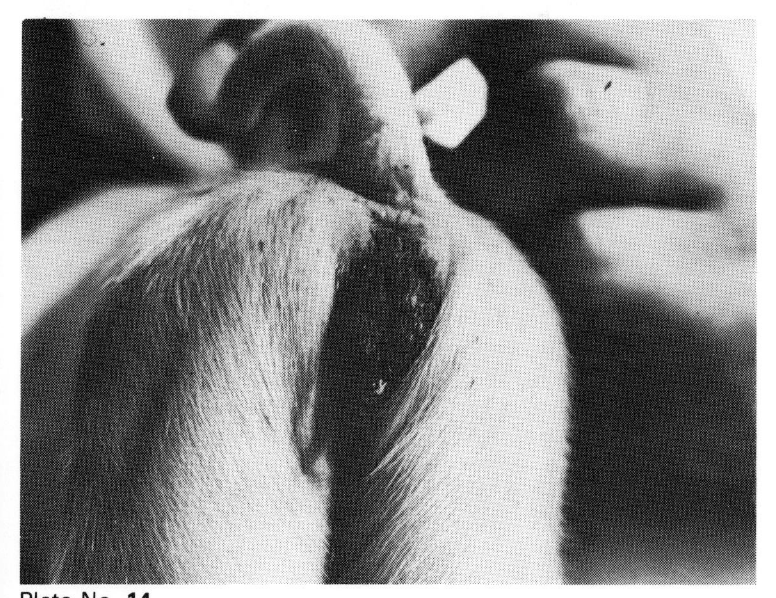

Plate No. **14**

Note bruised necrotic vulva.

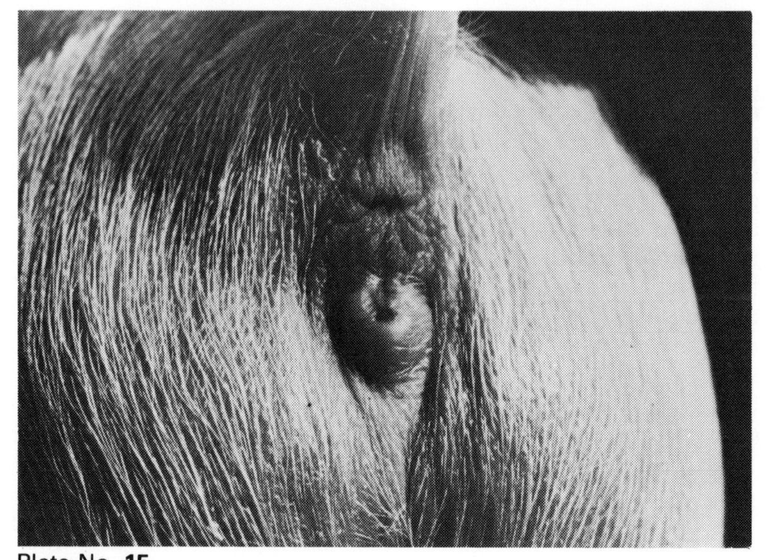

Plate No. **15**

Note small misshapen vulva.

COCCIDIOSIS

Coccidia organisms may be found in all ages of pigs. However, only the young pigs are likely to be clinically affected. The life-cycle of this protozoan parasite is complex and several species may be found in pigs but *Isospora* is probably the most pathogenic. Oocysts are voided in the dung of carrier animals and contaminate the environment.

These oocysts become infective in four to twelve days depending on conditions. Newly born piglets can therefore ingest infective oocysts from birth onwards. The oocyst undergoes a complex series of changes and multiplication in the gut of the piglet. During this multiplication stage severe damage to the epithelial cells of the gut wall may occur.

Clinical signs may appear as early as five days of age. Diarrhoea varies from a watery green colour to yellow and frothy with curd-like flecks and cottage-cheese-like consistency and affected pigs normally become stunted, hairy and emaciated.

The morbidity may be high but mortality is usually low. A feature of outbreaks is that litters are affected to a different degree and usually a few will show no adverse signs at all. The stockmen often complain of a musty foul odour throughout the whole farrowing house. There is no clear relationship between sow age and the degree of clinical infection of the piglets. Typically, the response to antibiotics is poor as the diarrhoea is a consequence of gut damage rather than infection. Treatment of clinically affected individual piglets with coccidiostats is often disappointing as the gut damage has already occurred by the time the disease becomes apparent. Control of the problem is best achieved with the use of in-feed coccidiostats to the dam from the time she enters the farrowing house until ten days post farrowing. Most of the common coccidiostats used in poultry feeds have proved successful in controlling the disease in pigs. Romensin premix (Momensin sodium Elanco) in the feed (1 kg/tonne) from five days before farrowing to ten days post farrowing has proved particularly effective in North-East Scotland.

TRANSMISSIBLE GASTRO-ENTERITIS (TGE)

This is a highly contagious enteric disease characterised by vomiting, severe diarrhoea and high mortality in piglets under two weeks of age. The diarrhoea is of a watery, greenish, foetid nature. As piglets get older they become less susceptible to the disease. Younger pigs which survive the disease often become runts. The incubation period is short (18 hours–3 days) and the disease rapidly spreads from litter to litter in a farrowing house.

Treatment
There is no specific treatment. Anti-serum from recovered sows given orally has had varied success. Piglets are more severely affected when milk is the only source of fluid. Earlier weaning may be beneficial in the face of an outbreak. Supplementary treatment with fluids and antibiotics to keep down secondary infections may be of some value.

Pattern of disease and control
The majority of outbreaks occur during the winter months. This may be linked to the susceptibility of the virus to sunlight and the winter movement of starlings. In England, the disease pattern has been characterised by a high rate of outbreaks over two years, followed by relative quiescence over the next four years. As herds have become larger and more intensive this pattern has become modified.

In many large herds the disease has taken a chronic low level persistent form. The outstanding feature in this type of herd is the continual source of susceptible pigs to perpetuate the infection. This occurs in herds with large farrowing houses with a continuous throughput and in those herds frequently purchasing gilts or store pigs for fattening. In these herds, piglets may not be affected until they have been moved to the follow-on accommodation.

Vaccines so far have not proved to be reliable, mainly because sows must themselves experience a *gut* infection with the virus before they pass on the antibodies *continually to the piglet via the colostrum* and *milk*. Good pig-flow within the unit and an 'all-in-all-out' policy in the farrowing house will often help to bring the disease under contol. If TGE has recently entered the herd and pregnant sows have not yet been exposed to the virus, two steps should be considered:

- Deliberate infection of those sows more than 2½ weeks from farrowing.
- For those animals which are due to farrow in 2½ weeks or less, attempts should be made to protect them from exposure by the creation of special facilities and managemental procedures until at least three weeks after farrowing, e.g., movement of sows in this category to another building or into arks away from the main unit.

Prevention

The disease is carried by pigs and other hosts contaminated with infected faeces, e.g., man, domestic pets and vermin. The steps taken to prevent entry of the disease should be obvious to any competent unit manager or owner. (TGE had not been diagnosed in Scotland at the time of writing.)

AUJESZKY'S DISEASE

This viral disease of pigs had become a scourge in many countries of the world and is beginning to assume worrying proportions in England. Known also as Pseudo-rabies (or mad itch in cattle), it is caused by a Herpes virus which is a very persistent infectious agent capable of infecting many hosts and spreading laterally from farm to farm, without the movement of pigs being involved, although most cases can be traced to movement of infected pigs and feeding of garbage. The disease is invariably fatal in cattle and domestic pets, i.e., they are dead-end hosts for the virus.

Clinical signs

All ages of pigs may be affected but the virus may remain dormant in a herd for long periods. Indeed infection without clinical symptoms was regarded as the norm in the USA for many years. When the disease becomes apparent the clinical signs vary with the age of the animal as follows:

Adult: Coughing, anorexia, depression, vomiting, constipation and high temperatures may be noted in sows and first-litter gilts. In addition, some may abort or eventually produce litters with dead mummified piglets. A few animals may develop nervous signs such as convulsive twitching and eventual prostration, and some of these may die.

Suckling piglet: Nervous symptoms with high mortality may be seen in young piglets. Vomiting, trembling, incoordination, convulsions, circling and backward movements are commonly seen. Temperature is raised (40·1°C, 105·5°F) and usually the outcome is death in newborn piglets but losses decrease to 40–60 per cent in four-week-old piglets.

Weaners/Fatteners: Coughing, fever and the gradual onset of nervous symptoms may be expected in a variable proportion of pigs in this group.

Control and prevention

There is no treatment for the disease with the exception perhaps of antiserum in some countries. The disease is notifiable in some countries (e.g. UK) but, at the time of writing (1982), no country has an eradication policy. Herd eradication may be attempted by blood testing breeding stock and removing positive animals. However, it should be borne in mind that recrudesence has occurred in some herds which had achieved clear blood tests for a number of months. Vaccination with both live and dead vaccines will reduce mortality considerably but will not prevent infection of animals with the field virus. Vaccination will, however, reduce mortality markedly in the breeding herd, but the disease may still be a problem in the fattening herd. The virus can exist in vaccinated herds and can spread from these herds to other herds. It is sometimes argued that the use of vaccines in breeding herds would prevent the identification of herds free from the disease as the results of blood tests would be confusing. However, if the use of vaccine is curtailed to breeding stock only, then this should not exclude any herd from achieving freedom from disease status, as the fattening pigs in the herd could be used to test for the presence of the disease (as disease sentinels). A sero conversion (positive blood test) in the fattening pigs would indicate the presence of the field virus itself. Vaccination of fattening pigs has met with varied success because of the high challenge by the virus when such large numbers of pigs are kept in close confinement. Irish workers at Stormont have reported best results with live attenuated vaccines in trials. An 'all-in-all-out' policy combined with vaccination would reduce the prevalence of the disease in fattening herds, but this can rarely be achieved in practice.

Breeding animals should only be bought from herds known to be free by regular testing. In spite of strict herd security, herds in areas of high pig density may still break down with the disease because of lateral spread by unknown means. Vermin are often blamed, but this has not been proved conclusively.

VOMITING AND WASTING DISEASE

This disease attacks piglets of one to three weeks of age and usually affects half the piglets in the litter; most of the litters at risk at any one time being affected.

Clinical signs

Affected piglets vomit, become constipated, gradually lose condition, becoming thin and hairy. Most affected piglets die in 1–2 weeks or have to be culled. Sometimes, nervous symptoms may be observed and sows have been said to abort.

Treatment

There is no known specific treatment. Supportive therapy with vitamins and fluid therapy may be useful.

Control and prevention

Purchase of gilts seems to be a precipitating factor. The virus is widespread in England and Northern Ireland and it is presumed that clinical disease only arises when carrier stock are introduced to a completely susceptible herd. No vaccine is available. Purchase of breeding stock should be kept to a minimum and should only be from reliable sources.

ROTAVIRUS INFECTION

Rotavirus infections have been demonstrated on a worldwide basis and may cause diarrhoea of suckling and weaned piglets.

Rotavirus may be the only agent present but is more often found in association with other pathogens such as coccidia and pathogenic *E. coli.* Affected piglets (3 to 35 days of age) may show depression of appetite after 36 to 72 hours and a profuse yellowish scour with particles of undigested milk. Mortality has been reported to be as high as 33 per cent. There is no specific treatment for the disease but fluid replacement therapy and weaning affected suckling pigs will help to control this diarrhoea. Transfer to a sow which has just reared a healthy litter may help. Fortunately, most outbreaks of Rotavirus infection cease within four to six weeks from the start of the outbreak.

Control
Milk from exposed sows has a high protective value according to research findings. Therefore, pregnant sows should be deliberately exposed to faeces from piglets suffering from the disease. Vaccines should be available in the near future.

CONGENITAL TREMOR TYPE A 11

This is a congenital disease of piglets due to infection of the foetuses with a virus. Clinically, it is characterised by rhythmic tremor of the head and limbs, reducing when the animal lies down and passing off when it is asleep. In piglets which survive long enough, the symptoms gradually disappear by two to three weeks. However, even when no longer obvious, tremor can be produced by a sudden unexpected stimulus, e.g., rough handling or sudden noise. Severely affected piglets look as if they are dancing on the floor. When the disease first enters a herd, 50 to 100 per cent of the litters may be affected. Mortality may also be high if artificial feeding or help with suckling is not given. An explosive outbreak is often seen after the purchase of a group of gilts which are carriers of the infective agent.

Once the disease has swept through the herd, future outbreaks are limited to a small percentage of gilt litters. Sows which have produced affected litters will produce normal litters thereafter if kept for rebreeding. The infective agent has not yet been isolated in the UK and the mode of transmission is not known. After an outbreak has begun, the herd should be closed and no further gilts brought in until the disease has almost died out. Do not dispose of females giving rise to affected litters. They will become immune and produce normal offspring the next time round. Young unbred animals should be exposed to affected piglets and cleanings before service takes place. There are other causes of congenital tremor (see earlier section on Genetic Disease).

ERYSIPELAS

This disease may take a number of forms in the pig. Infection may lead to sudden death (septicaemia), heart disease, arthritis, illness with high temperature and appearance of skin lesions (diamonds) (see Plate 16) and abortion or, more commonly, no disease at all. The causative organism (*Erysipelothrix insidiosa*) may be found in the soil, normal pigs, human beings and a wide variety of birds, mammals and fish. The various forms of the disease may be precipitated by certain factors, many of which are still unknown.

The following have been suggested by workers in different centres throughout the world—bad hygiene, atmospheric conditions such as hot, muggy weather, sudden change in diet, excessive weight gain, exposure to cold, lack of salt and vitamins in the diet, lack of natural resistance, and increased virulence of the organism. Perhaps these attempts at postulating a triggering factor can be summed up in the words of one worker, Dr Dubos: 'The state of susceptibility could change from day to day, indeed hour to hour, in response to all stimuli, physical, chemical, physiological or emotional—which constituted the total environment.'

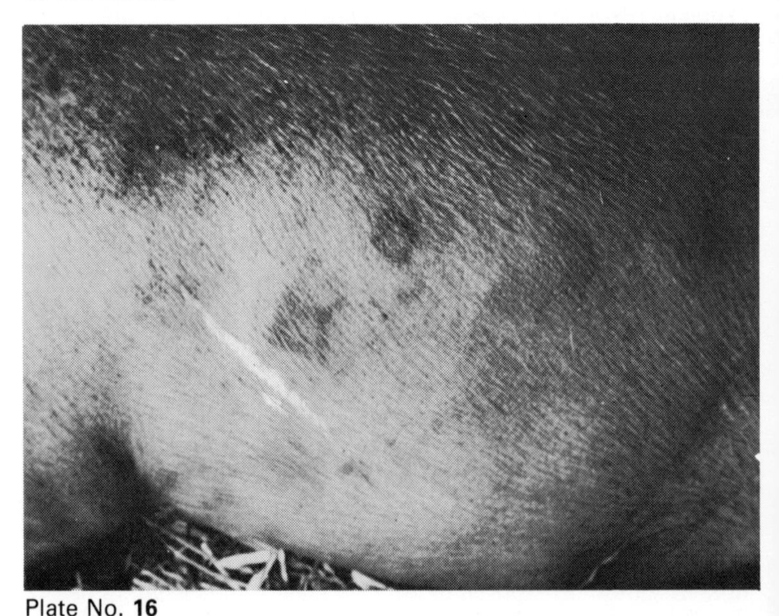

Plate No. **16**

Typical skin lesions of subacute erysipelas.

Diagnosis
Apart from the diamond form, a diagnosis of erysipelas infection can only be made by a veterinary surgeon—sometimes requiring complex and exhaustive laboratory tests. Isolation of *Erysipelothrix insidiosa* from arthritic joints of chronically affected gilts is not always easy.

Treatment
Both acute and sub-acute cases respond well to injections of penicillin. There is no evidence that the organism has become resistant to this drug. Chronically infected animals (especially arthritic and heart cases) show litle response to treatment of any kind and should be culled.

Prevention
Fortunately, erysipelas infection need no longer be a problem in the present-day pig unit as very effective vaccines exist (in the face of an outbreak antisera may be given to in-contact pigs; antiserum will have an immediate protective effect and last for at least 14 days). Some herds have never experienced a case of erysipelas, while others have found it to be a regularly recurring problem. For problem herds, protection by vaccination is effective, cheap and simple. Consult a veterinary surgeon about the timing and frequency of vaccination. Some vaccines may be safely given to pregnant sows while others are not recommended for this purpose.

EPIDEMIC DIARRHOEA TYPE I

This is an infectious enteric disease of sows (and finishing pigs) characterised by a profuse, smelly, greyish, watery scour, lack of appetite, normal or subnormal temperature and vomiting in a few cases. Occasionally, affected sows have been observed to abort. The disease may last for 3–8 days in sows and it rapidly spreads throughout the herd after introduction; 40 to 95 per cent of sows may be affected. The disease is not fatal and affected animals recover without treatment. However, it is regarded as a serious disease in lactating sows because it leads to agalactia. Piglets suckling affected sows do not normally scour but may die of starvation if not fed artificially.

The cause or mode of transmission is not known. Occasionally, it may be associated with the purchase of gilts but, more often, the herd may have been closed for some time when the disease appears. There is no specific treatment. Appetite-stimulating drugs and supportive vitamin therapy may help affected sows to recover more quickly. A veterinary surgeon should be consulted as this disease may be confused with swine dysentery.

Control measures

Attempts should be made to spread the disease within the dry sow population as soon as possible. This may be carried out by inoculating the food or the water with diarrhoeic faeces or vomit.

EPIDEMIC DIARRHOEA TYPE II

This disease was first reported in East Anglia in the spring of 1977. It differs from epidemic diarrhoea in that pigs of all ages may be affected. In this aspect, it is similar to TGE but pre-weaning mortality is low. Explosive outbreaks of diarrhoea have occurred, affecting all age groups and quickly spreading through affected herds. Affected animals show 'hosepipe' diarrhoea with light coloured faeces and some sows may vomit. In adults, the diarrhoea may last for three days but inappetence may continue for seven days. The disease is not fatal in adults or finishing pigs and 50–95 per cent of the herd may be affected. There is no specific treatment. No advice on control and prevention can be given until more is known about the cause and epidemiology of the disease.

SWINE DYSENTERY

Swine dysentery is a serious enteric disease of pigs which may affect pigs from three weeks of age upwards.

Aetiology

The infectious organism responsible is a spiral-shaped bacterium (*Treponema hyodysenteriae*) which causes severe inflammation of the large bowel. The organism is unusual in that it requires the presence of another organism or organisms which are probably found in the large gut of all normal pigs in order to cause the disease. Pure cultures of the organism will produce classic swine dysentery when fed to conventional pigs but will cause little or no symptoms in gnotobiotic pigs, i.e., pigs without organisms in the gut. The disease is usually spread from unit to unit by carrier pigs or indirectly via infected dung on boots or from lorries carrying infected pigs. It is not impossible that rodents and birds may act as mechanical carriers as well.

Clinical signs

Affected animals have a high temperature, appear dehydrated, staring coated, tucked up, restless and lose condition rapidly. They pass diarrhoeic faeces which usually contain varying amounts of blood, mucus, necrotic material and bile-stained undigested food. Another sign of this disease is depraved appetite. Affected pigs will drink urine and eat coarse rubbish rather than the fresh food usually available. They may die at any time during the course of the disease, recover completely, or become runts.

It has recently become apparent that the feeding of Emtryl and Virginiamycin as growth promoters may suppress the disease to such an extent that the classical signs of the disease are no longer present. Affected pigs have a blackish grey scour with a little mucus, no blood and an oily sheen. They may not lose condition but will certainly have a poor food conversion efficiency. Diagnosis, which can only be made by a veterinary surgeon, is made more difficult in these cases. In a similar way, breeding herds treating suckling pigs and weaners with Tylan-medicated creep may also, unknowingly, be harbouring swine dysentery in a sub-clinical form.

Treatment and control

Affected pigs may be treated by injection or medication of the feed or water. Medication of the water is the treatment of choice mainly because the ill pigs are unlikely to consume enough food to achieve the recommended intake of drug. However, pigs of 3–4 weeks of age should be treated by medication of the water and by injection, not only because their intake of food will be depressed, but also because the sows' milk provides part of their fluid intake.

Water should be medicated for at least 3–5 days and this should be followed by medicated feed for at least 5 days at therapeutic level. Your veterinary surgeon will prescribe the most suitable drug. It is important that the feed containing the drug should be thoroughly mixed and the drug incorporated at the correct level. Where wet-feeding systems are in use, the only effective method of treatment is to include a soluble drug at the correct level in the water fraction of total wet allocation.

OSTEOMALACIA

Osteomalacia is a disorder of the bone in the adult animal caused by a deficiency of calcium, and is now more frequently seen in the later stages of lactation, especially in first-litter gilts. The disorder becomes clinically apparent when spontaneous fractures of the bone occur. The pelvis and long bones are more frequently affected and posterior paralysis with a dog-sitting position may be noted. Occasionally a fracture is induced immediately after weaning when sows are mixed in groups or during service.

Precipitating factors may be:
Poor appetite in modern hybrid gilts; breeding a much younger and hence immature, rapidly growing female; the rearing of much larger litters by hybrid gilts; or high farrowing house temperatures. The problem may be prevented by increasing food intake, e.g. by wet feeding two to three times daily and by providing an adequate level of calcium and vitamin D in the diet. Some American researches have suggested the need for a higher level of calcium and phosphorus in the diet than that recommended by either the ARC or the NRC.

FOOT AND LEG CONDITIONS IN SOWS

It has been estimated by various research workers that foot and leg conditions are responsible for one-third of cullings in the breeding herd. There are several causes, such as infectious arthritis, accidents and some specific diseases, but perhaps one of the most common causes in present-day piggeries is the housing of sows on slats which have been improperly made or have markedly deteriorated. Rough, broken, chipped edges with excessive gaps is one of the main causes of lameness, leg and back weakness.

There are other factors such as inherited susceptibility (mainly Landrace and some lines of Large Whites) and the housing of maiden or newly-served gilts on slats of any kind. With regard to the latter, there is a higher incidence of culling during the first period of confinement and also later in life. The tendons, muscles, joints and bones of young maturing females require exercise on solid floors for proper development. Badly laid concrete, concrete with exposed sharp aggregate and excessively smooth concrete will also lead to various injuries.

Symptoms
Affected animals spend more time lying down. While standing,

they gradually develop excessive paddling movements of one or both hind feet. The soles of the claws become bruised and sore and the excessive paddling movements add insult to injury. The affected animals then attempt to remove the hind legs from the slatted area by arching the back and tucking the hind feet under the abdomen. At the same time, the lower limb becomes over-extended so that the hocks and the accessory digits come into contact with the slats (see Plate 17). The arched back causes pressure on the spinal nerves leading to partial or complete paralysis. Affected sows often sit with both hind legs extended straight forward.

Treatment, control and prevention

Affected sows should be removed to a liberally strawed pen as soon as symptoms appear. Some may have to be culled after farrowing. Maiden gilts should not be housed in slatted stalls until at least the last month of pregnancy. Properly made slats with rounded 'pencil' edges will prevent the problem occurring. The slats should be at least 100 mm wide. 127 mm slats with 25 mm gaps have proved to be very comfortable and clean. Perhaps the most technically advanced slat is the 'Triform' slat developed at the Scottish Farm Buildings Investigation Unit.

CYSTITIS/NEPHRITIS

Cystitis is inflammation of the bladder and nephritis is inflammation of the kidneys. In the sow, the most common cause of both is bacterial infection. The urethra leading from the sow's bladder to the vulva is relatively short and wide. In housing systems such as stalls, tethers and cubicles the vulva frequently becomes contaminated with faeces. This allows the many bacteria in the dung to enter the urethra and proliferate there, and finally they enter the bladder. In close confinement, sows *do not* urinate frequently and this would seem to be an important factor in the progression of the disease. As the urine is retained in the bladder the pH gradually changes to become more alkaline and this allows the organisms to multiply even more. Eventually, an inflammation of the bladder occurs and this may extend to the kidneys. Sows in the third trimester of pregnancy are most often affected. Clinical signs include increased frequency of urination, blood and pus in the urine and depressed appetite. The temperature is occasionally raised.

Prevention

Faeces should be removed from stalls and tethers once daily if time permits. Sows should be fit and not fat and given exercise where possible. Increased water intake can be stimulated by flooding the trough with water and sprinkling a little meal on the top. The idea of increasing water intake is to increase the rate of urination, so flushing out the bladder more often. This can also be achieved by increasing the salt content of the diet to almost 1 per cent, providing water is available ad libitum.

Treatment

Most sows will respond to daily injections of a broad-spectrum antibiotic for a four- to five-day period. However, once sows become clinically cured they nearly always relapse a few months later. Whole-herd treatment with penicillin (500 g/tonne feed for five days) every three months has also reduced the incidence of the problem to tolerable levels.

Affected sows should be culled in due course.

EXTERNAL PARASITES

MANGE

Mange may affect both sows and suckling pigs and is caused by a small parasitic mite called *Sarcoptes scabei var suis* (see Plate 18).

Life cycle

The mites live permanently on the pigs' skin and all forms (eggs, larvae, nymphs and adults) develop underneath the top skin layer in small burrows. Eggs laid in these burrows hatch within five days and develop quickly through the larval and nymphal stages to adults. The cycle from the egg to laying female only requires 10–15 days. One egg-laying female can be responsible for the production of several thousand mites within eight weeks. Transmission from pig to pig is by direct contact, the suckling pigs usually becoming infected from the dam. Although the whole life cycle is completed on the pig, mites may survive for 2–4 weeks if they drop off into a moist, warm area.

Clinical signs

Early lesions usually begin around the face, ears and any place where the skin is soft and tender. Much of the reaction is due to sensitisation of the pig to the mites. This causes intense irritation through localised and generalised body reaction in the form of

Plate No. **17**

Note abnormal posture and contact of accessory digits with slat.

Plate No. **18**

Adult and immature mange mites.

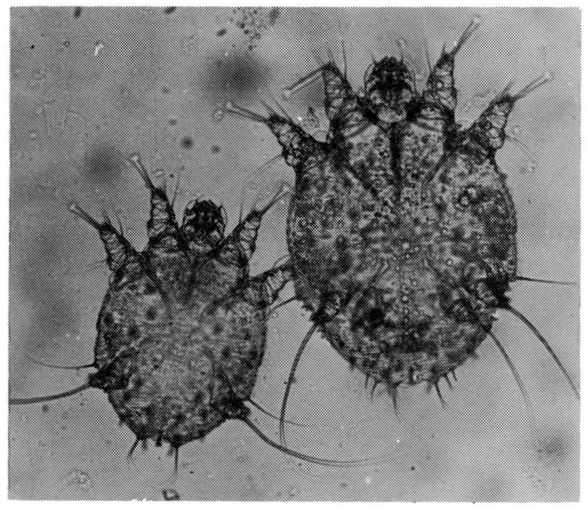

small red vesicles and papules. Gradually, the infested area becomes scaly, thickened, rough and dry. Eventually, the skin becomes thrown up into large thick crusty folds. The intense pruritus causes the pig to scratch and rub vigorously, thus damaging the skin and liberating serum which coagulates, dries and forms more crusts on the surface (see Plate 19).

Factors affecting severity of the disease

Many pigs will harbour mites throughout life without apparent harm or clinical signs. Malnutrition, enteric and systemic infections or debilitating factors appear to increase development of clinical lesions. Overstocking and any form of stress will also contribute to the severity of the lesions.

Effects of infestation

Pigs affected with pruritus have a poorer food conversion efficiency. The constant irritation causes pigs to become restless and often precipitates outbreaks of ear biting, tail biting and flank erosion. Although these conditions are usually seen in weaned pigs, they have also been noted in multi-suckled litters. Affected litters should be treated before weaning, especially in the case of early weaning into cages.

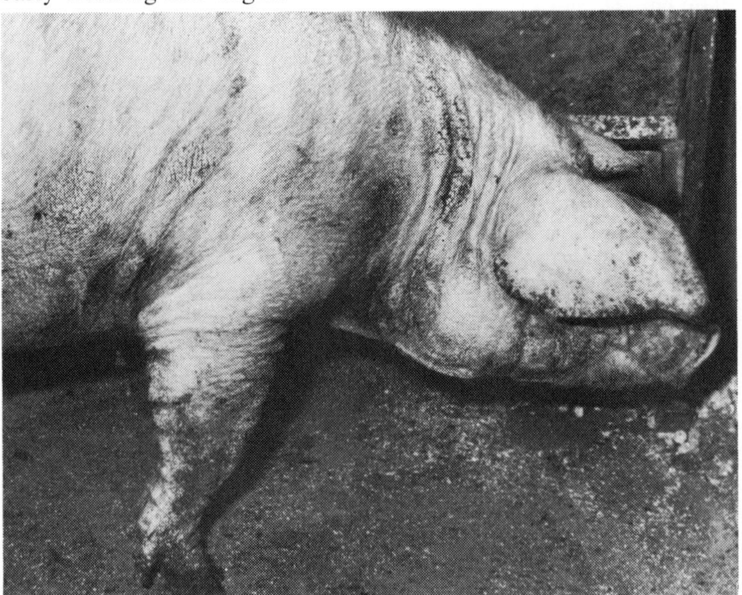

Plate No. **19**

Typical lesions of chronic mange infestation.

Diagnosis
The layman may easily confuse severe lesions of mange with exudative epidermititis, parakeratosis, some vitamin B deficiencies, and *dermatosis vegetans*. A veterinary surgeon should be consulted if in doubt.

Treatment
When mange is first diagnosed in a herd, all sows and litters should be treated with a suitable preparation at least twice at an interval of 10–15 days. The second treatment is essential as eggs may hatch out after the first treatment. Spraying severely affected animals is unsuccessful in many cases. Sows should be scrubbed throughly with mild soapy water before the mange wash is applied. The insides of the ears must always be treated as this is a common source of infection. Once the breeding herd has been treated twice, it is then only necessary to treat sows once before they are taken into the farrowing accommodation.

LICE

The pig louse *Haematopinus suis* spends its entire life on the pig. The female attaches the small white eggs to the hairs of the pig and the eggs are easily visible to the naked eye. The entire cycle from egg to egg laying adult takes 20–30 days. Transmission is by direct contact from pig to pig. When dislodged from the pig the lice can only survive 2–3 days on the ground or even on humans. They are capable of feeding on the latter during this short period. Lice feed on the pig by piercing the skin and sucking the serum and blood.

Clinical signs
The piercing of the skin causes mild irritation and pruritus which the pig attempts to relieve by scratching. This leads to restlessness and poor food conversion efficiency. Heavily infested suckling piglets may become anaemic. Lice may also act as vectors of pig pox and other viruses.

Treatment
The treatment suggested for mange will effectively eradicate pig lice from the herd.

INTERNAL PARASITES

Internal parasitic infections may be divided into three groups—lungworms, bowel worms and stomach worms.

LUNGWORMS

These cause parasitic pneumonia and are only found in pigs with access to soil as the infective larvae can only be picked up by eating earthworms in which they develop. Persistent coughing and loss of condition in sows housed in paddocks may be due to lungworm infestation.

STOMACH WORMS

These cause loss of condition and anaemia in sows of any age. The life cycle is direct and uncomplicated. Heavy infestation may occur in paddocks or strawed yards.

BOWEL WORMS

There are a number of bowel worms varying from the large Ascaris worms to very small worms which may burrow into the wall of the intestine. Severe infestation may lead to a variety of signs—loss of condition, diarrhoea and sometimes dysentery. In addition, heavy infestation with Ascaris will cause losses through condemnation of livers through which the larval forms of these worms migrate after ingestion of infective eggs.

Treatment

There are many drugs on the market which will effectively remove all the above types of worms. Sows should be dosed before farrowing while young pigs should be dosed just after weaning. Strict attention should be paid to instructions on the leaflet with the drug. Overdosing with Haloxon (Whithelmin, Loxon) may cause posterior paralysis three weeks later. Consult a veterinary surgeon about treatment and best time to dose in your own unit.

Generally speaking, worms no longer cause serious problems in the large modern intensive units of today—mainly due to the widespread use of slatted or perforated floors and the plentiful supply of reasonably priced, effective drugs.

A negative worm egg count does not mean that a pig is worm-free. The number of eggs counted in faeces samples cannot be directly related to the number of worms present in the bowel so that egg counts are not necessarily a good indication of worm burdens. Sows reach their own level of immunity, i.e., they will tolerate a certain number of worms in the gut and this level will not rise in spite of exposure to infective eggs or larvae.

A sow harbouring 2,000 worms before dosing will eventually harbour the same number again. The egg of the Ascaris worm may live for many years in hostile conditions. It is often the first parasite to gain entry to minimal disease or specific pathogen-free herds. Egg production rises rapidly just after farrowing. This is one reason why sows should be dosed before farrowing.

STRESS

The word stress has been used with gay abandon to cover a multitude of situations which have either been given little thought or proved too complex for a simple explanation.

The term stress, at least in the context of this book, is used as defined by Seleye in 1950, who states that 'the term *stress* be used to denote the reaction or response of the animal to adverse influences or stimuli and that the actual stressful influences, and stimuli themselves should be referred to as *stressors*'. There are many stressors, both physical and social, which may provoke short-term specific and non-specific reactions within the body of the animal. These changes within the body systems result in physiological and behavioural aberrations. In a sense, stress as it is known in the pig industry is a non-specific long-term response by the pig. The pig attempts to resist, or adapt to, one or a combination of stressors in order to maintain equilibrium within the body systems (homeostasis). In the young piglet, such stressors have the same effect on the gastro-intstinal tract (the gut), namely, slowing of gut contractions, especially the stomach, congestion of blood vessels, haemorrhages of its lining and pooling of blood in the abdominal vessels. This stimulus response mechanism is delicate and sensitive. When stressors become overpowering, there is no doubt that over-compensation can occur in the body, resulting in as harmful effects as the stressor itself.

Stressors of short duration and small intensity play a vital role in the normal development of the pig, and in the long term are probably beneficial to its survival. However, pigs no longer live in their natural environment and are continually exposed to many stressors, both social and physical. It is not surprising that some should succumb through sheer exhaustion of some body systems or over-react with consequent deleterious effects. Animals which are totally sheltered from outside or abnoxious stimuli may be said to be *under*stressed while those exposed to intense prolonged, over-stimulation by stressors are said to be *over*stressed.

A stressing agent first reacts in the brain through the normal senses of smell, touch, sight and hearing which, in turn, stimulate the brain to produce certain hormones which act on the body as a whole and also trigger off certain neuro-hormonal reactions in the gut and other body systems. One of the main hormones released belongs to the steroid group and is responsible for a reduction in resistance to infection, increased breakdown of body protein and interference with the utilisation of certain nutrients. Overstressed pigs are more likely to succumb to infectious diseases and exhibit behavioural aberrations.

Supplementary reading

Diseases of Swine, Iowa State Press, Ed. A. D. LEMAN.

Chapter 4

REPLACEMENT POLICY

THE EXPENSE of stock replacement depends on the prevailing cost of the replacement in relation to the value of the culled animal. Replacement policy must be considered very carefully as the decisions made can have far-reaching effects on the health, performance and profitability of the herd. The replacement rate depends on the breeding and replacement policy and on the wastage rate.

WASTAGE RATE

On average, sows in commercial herds last for between three and four litters before culling or death, the wastage rate in successive parities for one comprehensive study being as shown in Table 4·1.

TABLE 4·1. Wastage in gilts and sows

Number of litters at time of culling or death	*Proportion of sows*
0	10
1	22
2	12
3	12
4	7
5	10
6	10
6+	17
	Total 100

Source: Jones, J. E. T., London Veterinary School

It can be seen that 10 per cent of sows failed to produce even one litter, 56 per cent of sows failed to produce more than three litters, while only 37 per cent of sows produced more than four litters.

CAUSES OF WASTAGE

Causes of wastage determined in various surveys are roughly as indicated in Table 4·2.

TABLE 4·2. Causes of sow wastage

Causes of wastage	*Proportion of sows (%)*
Reproductive failure	32
Locomotor disturbances (lameness and paralysis)	32
Udder problems and milk shortage	5
Old age	8
Death	12
Miscellaneous (savaging, clumsiness, small litters, poor performance and grading of progeny, adverse economic situation etc)	11
Total	100

Source: Jones, J. E. T., London Veterinary School

Thus, death was responsible for about 12 per cent of sow wastage, and only about eight per cent of sows were culled because of old age. The remainder were culled because of disease, other defects or poor productivity before they attained the age at which producers would normally cull them. It can be seen that the major causes of culling were breeding problems and locomotor disturbances (lameness and paralysis) which each accounted for about one-third of the wastage, so that efforts to reduce wastage rate in sows must concentrate primarily on these two problems.

Culling rate of both sows and boars will also be much higher in herds in which a policy of rapid generation turnover, linked to a soundly-based genetic improvement scheme, is being followed in order to increase rate of genetic improvement within the herd. Culling rates will also increase during times of low profitability and when the value of the cull sow is high relative to the cost of a replacement gilt.

SOW PERFORMANCE ACCORDING TO LITTER NUMBER

How important is it that such a high proportion of sows fail to produce more than four litters? Obviously it is important to minimise the incidence of death, agalactia, lameness, low litter size, breeding difficulties and similar problems which lead to early culling. Moreover, enforced culling imposes a severe restriction on the purposeful culling and selection of animals which is necessary for continued genetic improvement in terms of economically important characteristics. But are there other real losses involved in the fact that such a low proportion of sows fail to produce more than four litters?

In order to answer this question one must examine the relationship between litter number (parity) and sow and piglet performance.

NUMBERS BORN AND REARED

The relationship between parity and number of piglets per litter which were stillborn, liveborn and reared to three weeks of age is indicated in Figure 4·1.

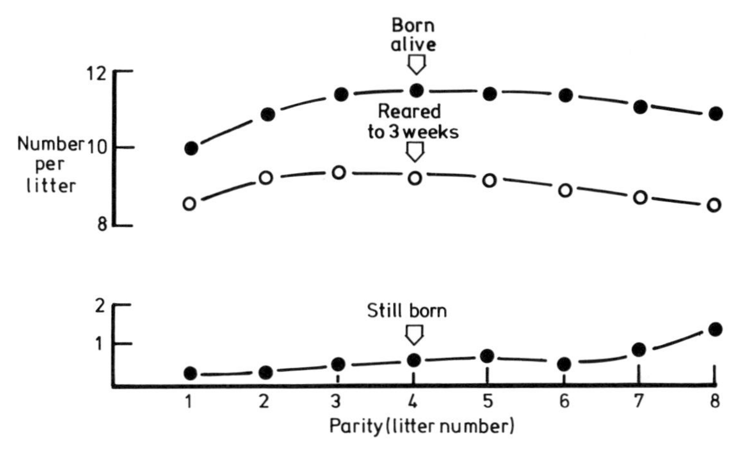

Figure 4·1. Relationship between parity and numbers born and reared per litter.

*Source: Strang, G. (1968). Analysis of PIDA Litter Records.

Thus, on the basis of the data in Figure 4·1, there tends to be a fairly sharp increase in incidence of stillbirths and a decline in number born alive after the sixth litter. Number of piglets reared declines after the third litter and this decline accelerates after the fifth litter. There also tends to be increased variation in weaning weights after the fifth litter.

There is an increase in losses of piglets after the third litter. This is likely to reflect the decreasing efficiency of the sow after this stage. With advancing litter numbers, mean birth weight of piglets declines, variation in piglet birth weights increases and the incidence of such problems as clumsiness and agalactia also increases. All these factors contribute to greater piglet losses with advancing parity.

WEANING-TO-CONCEPTION INTERVAL

Weaning-to-conception interval is also influenced by litter number in that this interval is longest after the first litter. In successive litters, thereafter, this interval is fairly constant. (See Chapter 10, Figure 10·4, Page 264.)

MAINTENANCE REQUIREMENT

Sows continue to grow, including skeleton and muscles, up to the fifth or sixth litter. For the sow to stay in reasonable body condition with advancing age, she should gain weight from litter to litter. A weight gain from one litter to the next of 12–15 kg will maintain most sows at the desirable level of body condition; this level of gain can be taken as an indicator that feeding level is pitched at the right level, i.e. the sow is neither being under nor overfed. Consequently, if feeding practice is correct, the sow will increase gradually in weight up to the fifth or sixth litter. Because the amount of food required purely for maintenance of normal body functions increases with liveweight, the older, heavier sow will have a higher food requirement than the younger sow. The increase in food requirements as the sow becomes older and heavier is indicated in Figure 4·2.

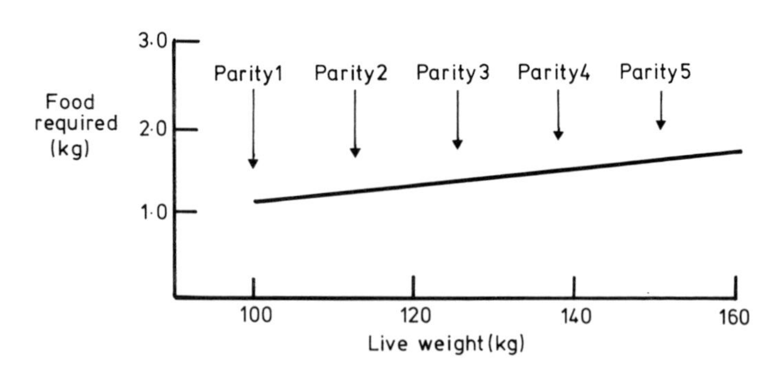

Figure 4·2. Food required for maintenance at 20°C.
(Diet with 12·7 MJ. Digestible Energy per kilogram).

Source: Hovell, F. D. De B., Gordon, J. A. and MacPherson, R. M. (1977). Rowett Research Institute.

Thus, older sows with higher food intakes need to be more productive to cover the extra costs. While sows are most productive between litters two and five, they tend to become less efficient thereafter because of their higher food requirements and declining production. Some producers cull sows routinely after the sixth litter, although individual sows can continue to be highly productive and efficient after this stage. A policy of culling sows only after they have produced a poor litter, means the sows have been retained for one litter too many. On balance, in order to achieve both faster genetic improvement and more efficient use of food, it is desirable to avoid retaining too many old sows in the herd.

In view of the fact that sow productivity is maintained at a high level up to about the fifth litter, it is somewhat unfortunate that such a high proportion has to be culled before this stage because of lameness, failure to breed and other problems.

REPLACEMENT POLICY

The decisions to be made about replacement policy should be based on:

- Whether a closed or open herd policy is to be followed.
- Health considerations when purchasing replacements.
- Rate of stock turnover, i.e. replacement rate.
- Basis of selection or purchase of replacements.
- Care in introducing purchased replacements to the herd.

A CLOSED OR OPEN HERD?

There are certain advantages to be gained by a closed herd policy. In this context, a 'closed herd' is defined as one which breeds its own replacements and introduces fresh genes in the form of semen only.

Advantages

The advantages to be gained are mainly on the health side and, of course, this is reflected in the overall herd performance. Infectious diseases, for example swine dysentery, will be conspicuous by their absence in a closed herd with only such rare exceptions as infections carried by vermin or wind. In addition, 'managemental' diseases such as colibacillosis, abortion and the SMEDI syndrome will not be a problem. One herd of 120 sows (mainly first crosses but with a small purebred nucleus of both Large White and Landrace) in the North East of Scotland provides a good example of what can be achieved with a closed herd policy. The herd has been closed since 1963 except for the introduction of semen from top sires through AI. The performance figures from 1979 to 1981 are as follows:

Average number born/sow/year	24·9
Average number alive per litter	11·4
Average number weaned (6 weeks)	10·0
Litters/sow/year	2·17
Number weaners/sow/year	21·7
Conception rate to first service	94·5%
Pre-weaning mortality	12·3%
Post-weaning mortality	1·6%
Days to bacon (at 90 kg liveweight)	165
Food to liveweight gain ratio (weaning to bacon weight)	3·0

N.B. All figures include gilt litters.

Colibacillosis, rhinitis, swine dysentery, SMEDI syndrome, pig pox or epidermitis have not been diagnosed in the above herd in the last 10 years. Pneumonia (the herd is not enzootic pneumonia free) is not a problem in the finishing house. Fertility has been exceptionally good but this may also be a reflection on the high level of stockmanship and management prevailing on the unit. The 'disease-free' status achieved by a closed herd policy is due in part to a high level of herd immunity (especially with regard to such diseases as colibacillosis and SMEDI syndrome) and, in part, to the non-introduction of disease. Because of high herd immunity, the pens in the farrowing house and fattening house are never washed out or disinfected.

A partially closed herd policy may be achieved by breeding the replacement gilts and only buying in an occasional boar. However, even the introduction of a boar will bring in fresh strains of *E coli*, enteroviruses, respiratory organisms and sometimes more serious pathogens such as swine dysentery or TGE.

Disadvantages

The disadvantages of running a closed herd relate mainly to the breeding policy. Purebred lines have to be kept for the production of crossbred gilts. Apart from financial considerations, this will entail greater management input as more detailed records will have to be kept and some form of objective gilt testing and selection will have to be carried out. Extra capital will be required for housing the replacement stock. Great care must be taken in the selection of home-bred boars and more boars than necessary will have to be reared. Boar testing will require more capital and labour. A stockman will also have to become very efficient at AI.

In summary, then, the health status of a closed herd will be high in all respects and this will be reflected in good production figures. However, maintaining a closed herd requires a higher management input, and unless this input is provided, the genetic value of the herd will probably be less. An open herd policy will result in a greater health risk. Ways of minimising this risk are covered later in this chapter.

HEALTH CONSIDERATIONS WHEN PURCHASING REPLACEMENTS

Whether starting a new herd, expanding an existing one or merely replacing redundant stock, health considerations are of paramount importance. In any herd, protection of the health status of the existing stock takes priority but the protection of the health of the replacement stock must also be considered. Because of the substantial capital tied up in the present-day pig units and the low profit margins, owners should think carefully before purchasing replacements. The sources of purchased replacement stock may be divided into five broad categories, namely:

- Stock from specific-pathogen-free herds (SPF).
- Stock from minimal disease herds (MD).
- Stock from Pig Health Scheme herds (PHS).
- Stock from other sources (varying from some large breeding organisations to small herds).
- Stock from within existing herd.

SPF herds

SPF stock is usually produced by hysterectomy or hysterotomy and reared in strict isolation from other stock. Breeders advertising stock as SPF should be able to supply a written guarantee stating the pathogens from which the stock is free. Buyers should insist on a written guarantee if contemplating the purchase of stock thus labelled. This now comes under the Trades Description Act which should protect buyers to some extent.

MD stock

MD pigs are also produced initially by hysterectomy or hysterotomy and reared under the strictest quarantine conditions. MD is a relative term—it does not mean that such pigs will be disease-free. Recently, new MD herds have been established by the process known as Medicated Early Weaning (MEW) (Pig Improvement Co., Oxford, England). MD pigs should be free of most respiratory pathogens, swine dysentery, transmissible gastroenteritis, lice, mange, worms and all notifiable diseases. Some breeding companies may guarantee freedom from some of these diseases. However, it is wise to read carefully the small print in the relevant brochure.

PHS stock

Breeders who are members of the Pig Health Scheme in the UK receive free regular visits from both ministry and private veterinarians. A thorough and rigorous inspection of the whole herd is made at each visit. The aim of these inspections is to ensure a high health status within each herd. Should an infectious disease be diagnosed or suspected, the unit in question would not be allowed to sell gilts or boars until such time as the authorities were satisfied that the problem was resolved. This scheme helps to ensure that members' herds have a high health status.

Other sources

This category includes all other suppliers. Sources vary from large company breeding organisations to small open 'back street' breeders. Some large breeding companies have their own private health schemes run either by company-employed veterinarians or by consultant veterinarians from private practice. Other breeding organisations do not employ veterinarians on a regular basis but this does not mean that their stock will be affected with disease. Many smaller units now have routine veterinary inspections at regular intervals and have excellent health records. There is a wide variety of source and choice. It is best to consult your veterinarian before purchasing replacements. It is difficult and sometimes impossible to write certificates of freedom from some diseases because there may be no specific test available for these, e.g. infectious atrophic rhinitis. However, there is nothing to stop the buyer asking relevant questions, such as, has it been necessary to treat the herd in question for rhinitis or swine dysentery? Having decided to buy replacement stock it is wise to purchase from one source only. Avoid buying in-pig gilts or gilts from market sales.

REPLACEMENT RATE

In breeders' herds where a soundly based breeding improvement policy is being followed, there is merit in replacing both boars and sows fairly quickly since, if the improvement scheme is effective, the new generation of pigs should be superior to the existing stock. Thus in nucleus breeding herds, boars may be used only for about one year and sows only for two litters before being replaced. However, in commercial herds, it is customary and prudent to retain both boars and sows as long as they are fit and healthy and both they and their progeny are performing well. Exceptions to this general rule arise on occasion in relation to sows, as some producers cull routinely after the sixth litter since productivity and efficiency tend to decline after this stage. This policy, inevitably, involves disposing of some sows which would continue to be efficient for a few more litters. However, the alternative policy is to delay culling until the sow produces a poor litter and, in hindsight, this also has its snags since such a sow has been kept for one litter too many.

On balance, it is preferable to cull after a given number of litters, probably the sixth, although the exact stage may differ between herds. In some herds, there may be an accelerated decline in productivity and efficiency after the fifth litter, whereas such a decline might not set in until after the seventh litter in other herds. Consequently, an examination of the herd records will be necessary before a decision can be taken regarding optimum stage or parity at which to cull.

It is worthy of note, however, that whatever this stage is found to be, it will apply to a relatively small proportion of sows since most culling and wastage will have taken place before this stage, for a variety of reasons (see Table 4·2) in most herds.

Whatever the reason for culling, it is vitally important that problems are detected promptly and culling carried out without delay. Even though breeding problems have been one of the major causes of sow wastage, this need not result in appreciable loss provided problems such as anoestrus are detected promptly and the sow quickly culled. Pregnancy diagnosis carried out four to five weeks after service is a very useful way of detecting 'passengers' and eliminating these before they consume too much food and unnecessarily occupy facilities. Individual boar and sow record cards should be kept up to date, examined frequently, and prompt culling carried out whenever problems present themselves.

ADVANTAGES OF CROSSBREDS OVER PUREBREDS

Whether replacements are home bred or purchased, it is desirable that commercial producers concentrate on crossbreds or hybrids for a variety of reasons.

The crossbred has the following advantages over the average of the two parent purebreds:

- Younger and lighter at first heat.
- Improved breeding regularity.
- Fewer embryonic losses.
- Higher numbers born.
- More uniform birthweights.
- Greater viability of piglets, therefore improved survival.
- Higher milk production.
- Larger litters at weaning.
- Higher total litter weight at weaning.

The advantages of crossbreeding in terms of litter productivity are illustrated in Table 4·3.

TABLE 4·3. Advantage of first-cross litters and back-cross litters from first-cross sows over purebred litters

	Number born per litter	*Number weaned per litter*	*Litter weight at three weeks*
First cross	+2%	+5%	+5%
Back cross	+5%	+8%	+8%

Source: Smith, C. and King, J. W., ABRO, Edinburgh. Analysis of PIDA litter records involving mainly the Landrace and Large White breeds and their crosses.

It can be seen from Table 4·3 that a litter out of a crossbred sow has about 8 per cent more pigs at three weeks of age and weighs about 8 per cent more at the same stage, relative to the average purebred litter of the two parent breeds. A litter out of a purebred sow and a boar of another breed is also slightly superior, on average, to a purebred litter, i.e. a five per cent advantage both in terms of number of pigs per litter and litter weight at weaning.

SELECTION OF REPLACEMENTS

Only in large herds can a completely closed herd policy be followed without involving a degree of inbreeding. Artificial insemination is very useful in such herds to help minimise inbreeding and as a basis for producing replacement boars and gilts since most boars used in AI are of high genetic merit as judged by performance test results.

Among the criteria which are useful to consider in selection of replacements are the following:

- Efficiency of sire, e.g. performance test results and/or records of progeny (i.e. progeny test data).
- Efficiency of dam as judged by records of previous progeny.
- Weight for age.
- Food conversion efficiency (involves individual feeding and recording food intake).
- Carcase fatness, e.g. as measured ultrasonically.
- General soundness. Normal anatomy and sound in feet and legs, in particular.
- Underline, i.e. abnormal teats (blind, inverted, damaged and extremely small teats), number of normal functional teats, evenness of spacing between normal teats.

Since breeding regularity and litter productivity have a low heritability, these criteria do not merit a great deal of attention in selection. These traits, being much more under the influence of management and husbandry, are best improved by paying attention to these factors.

A useful way to set about selection is first of all to preselect the litters from which replacements are to be selected on the basis of the proven breeding merit (available records on previous progeny) of the sire and dam.

Having preselected the litters from the best sires (either through AI or natural service) and from the 'elite' sows, these pigs should be grown under normal commercial conditions and tested near market weight.

A useful basis for such a test is a combination of weight for age and level of fatness as measured by ultrasonics since both of these traits are economically very important and are quite highly heritable. These two traits are incorporated into one *index* of merit in the 'On Farm Testing Service' run by the Meat and Livestock Commission in UK. The gilts with the highest index rating on the basis of these two traits in any batch of gilts tested are potentially useful herd replacements but their soundness, particularly of legs, must be checked, since most sows have to spend their life on concrete and thus sound legs are vitally important. In addition, all replacement gilts should have a minimum of fourteen functional, evenly-spaced teats.

If gilts with a high index rating (weight for age and carcase leanness) are either unsound on their legs or do not come up to the required standards for underline (number of functional teats and evenness of spacing), then they are unsuitable candidates for replacements. Those gilts with the highest index rating which are sound on legs and come up to standard on 'underline' should be selected as replacements.

If boars are also to be selected from within the herd, they should be selected using similar criteria to those used for gilts.

In a few commercial herds, individual feeders have been provided so that the food intake and food conversion efficiency can be monitored in prospective breeding boars. Such facilities tend to be expensive and entail high running costs. However, the information on food conversion efficiency provides a further, very useful basis for selection. It will be much more cost effective to confine such testing facilities to boars because of the very much larger influence a boar has relative to a gilt in a breeding herd.

For those commercial herds which breed their own replacement gilts and purchase boars, the 'criss-cross' breeding method is one of the most effective in keeping the female stock as crossbred as possible so as to maintain hybrid vigour and avoid inbreeding.

An example of the way such a system could be organised is as follows:

The foundation stock females could be first crosses between Landrace (LR) and Large White (LW). These would be numerically tagged with red tags. Only LW boars would be used on these females and these boars would also be identified with red tags. Selection of replacements from this cross would be on the basis of the criteria discussed earlier.

These selected progeny (three-quarters LW) should be tagged numerically with blue tags and only LR boars, also identified with blue tags, should be used to serve these females. The progeny from this mating will be predominantly LR (62·5 per cent); these should be given red tags and mated to LW ('Red' boars). So the criss-crossing, or alternative use of LW and LR boars in successive generations proceeds as indicated in Table 4·4.

TABLE 4·4. An example of criss-cross breeding using the Large White and Landrace breeds

Generation No.	*Colour code*	*Boar breed*	*Female crossbred*	*Percentage of LW in female crossbred*
1	Red	LW ×	A (LW × LR)	50
2	Blue	LR ×	B	75
3	Red	LW ×	C	37·5
4	Blue	LR ×	D	68·75
5	Red	LW ×	E	34·8
6	Blue	LR ×	F	67·4
7	Red	LW ×	G	33·7
8	Blue	LR ×	H	66·8
9	Red	LW ×	I	33·4
10	Blue	LR ×	J	66·7
11	Red	LW ×	K	33·3
12	Blue	LR ×	L	66·7
		etc		

The tagging system simplifies breeding, 'Red' sows (predominantly Landrace) always being mated to Large White ('Red') boars and 'Blue' sows (predominantly Large White) always being mated to Landrace ('Blue') boars.

As well as helping to maintain hybrid vigour, continuous improvement in economically important characteristics can be achieved through careful objective selection of replacement gilts and purchase of high-quality boars.

A system which is gaining in popularity in practice is the mating of first-cross gilts, e.g. Large White (LW) × Landrace (L) to first-cross LW × L boars. While this approach may result in a slightly lower advantage from heterosis relative to the mating of a purebred boar (e.g. LW to a LW × L sow), this appears to be its only slight disadvantage and it has the great merit of being simple to operate. It is vital, of course, that home-produced crossbred gilts are selected on a sound basis as previously described. The first cross LW × L boar tends to have greater libido, on average, than purebred boars. For these reasons, breeding companies are finding that demand is increasing for first-cross boars to be mated to home-produced crossbred gilts.

PURCHASING REPLACEMENTS

When buying in replacements, one must pay as much regard to the health status of the herd from which purchases are being contemplated as to the genetic merit of the stock. The criteria which can be used as a basis for deciding on where best to purchase replacement stock include:

- Previous experience of replacements purchased from a particular source.
- The claimed health status of the herd, supported where possible by confirmation of this by a veterinary surgeon.
- Performance and progeny test results on as large a sample of stock as possible from a particular herd or breeding company.

The Commercial Product Evaluation (CPE) Test, as organised by the Meat and Livestock Commission in the UK, is equivalent to a useful 'Which?' test on a variety of commercial crossbreds or 'hybrids' available from the largest commercial breeding companies. It involves the evaluation at a central station of a random sample of the products (boars and gilts) of commercial companies. Litter productivity up to weaning is recorded and the performance of the pigs on two feeding systems is monitored up to each of three slaughter weights. The physical performance is translated into economic terms so that the real commercial value of the sample of stock is determined and the products of the various companies can be compared on this basis.

So far the CPE has provided some useful guidelines regarding the relative merits of various commercial crossbreds or hybrids available to the farmer although a much larger sample of stock from the various companies would require to be tested in order to increase the reliability of the comparisons for the reproductive traits.

While such tests provide some basis for assessing the general efficiency of a company's pigs, the farmer often requires information on individual boars, in particular, and gilts as a basis for selection and purchase. An index rating is normally available on each pig available for sale. This index rating may be based on weight for age and carcase leanness (as measured by ultrasonics) assessment but sometimes, in addition, a food conversion figure is available on boars, which increases the usefulness of the index rating. Needless to say, sound legs and an adequate 'underline' are as important in purchased as in home-bred replacements.

A very detailed assessment of the performance of boars is also carried out by the Meat and Livestock Commission at several central testing stations in the UK. The assessment involves weight for age, food conversion efficiency and carcase traits combined into a single index of merit. Since the assessment is based very firmly on economically important traits, it provides a very useful basis for selection and purchase of boars.

CARE IN INTRODUCING PURCHASED REPLACEMENTS TO THE HERD

If practicable, bought-in stock should be kept in strict isolation for at least three weeks—preferably in a unit away from the main piggery. The purchased stock should be examined carefully for evidence of disease. It may be wise to ask for a veterinary examination. Gilts may be carriers of some diseases for which there are no specific tests. A useful on-the-farm test is to expose bacon pigs deliberately to the gilts still within the isolation unit. This may detect diseases such as swine dysentery and enzootic pneumonia (if your herd is already clear of EP) but even this test cannot be guaranteed to be 100 per cent successful.

Having reached the decision to introduce the gilts to the main herd, further steps will have to be taken to immunise them against the indigenous flora of the unit (see fertility problems, Chapter 3). They should also be injected with vaccines currently used in the herd.

SUMMARY

There is a high wastage rate in sows and the majority fail to produce more than three litters. The main causes of wastage are reproductive difficulties and lameness. The high wastage rate necessitates a high replacement rate and rapid turnover of stock which is useful for genetic improvement when a sound breeding improvement policy is being followed, but on the debit side it results in failure to exploit the potential of maturing stock.

Sows are most productive from the second to the fifth litter but, thereafter in most herds, productivity and efficiency tend to decline.

Replacement costs vary largely according to the relative cost of the replacement and the value of the cull. Replacements, whether home bred or purchased, should be selected on an objective basis, the most important criteria being weight for age, carcase leanness, underline and soundness of feet and legs.

It is vital that every precaution is exercised when purchasing stock to avoid bringing disease into the herd and replacements should be exposed to older breeding stock well before mating in order to provide them with an early opportunity to built up immunity to prevailing infection.

Chapter 5

MANAGEMENT OF THE GILT

THE GILT can often be a neglected entity on the pig unit. She is looked upon as an unproductive animal and is often treated as such. This attitude can have several repercussions. For instance:

- Gilts can be bred at too late a stage, resulting in excessive age and weight at first farrowing, thus increasing the cost of rearing. Such gilts will tend to be larger than necessary throughout their lives and the increased food required to maintain these large animals is unlikely to be covered by higher productivity.
- Gilts have, on average, much fewer piglets born than older sows. Many producers are very complacent about this situation and make no serious attempt to increase litter size but, since sows have fewer than four litters, on average, gilt litters constitute more than one-quarter of those in the UK National herd. Thus, small gilt litters can have a major effect in depressing the herd performance of numbers weaned per sow and gilt.
- Because of uncertainty as to when gilts will reach puberty (first heat accompanied by ovulation), and because some may turn out to be infertile, often a considerable margin of safety is allowed for in the number of gilts retained for breeding. This same uncertainty regarding when gilts will start cycling encourages a negative rather than a positive culling policy, i.e. many sows which should be culled because of performance defects are retained because of uncertainty over when replacement gilts will breed. The alternative is to cull such suspect animals in good time and avoid the risk of a poor litter next time but, in doing this, expensive facilities may end up not being fully utilised if gilts cannot be bred when required. Empty spaces in expensive accommodation will result in a fall in margins and make it more difficult for producers to meet high interest charges on buildings and other costs.

OBJECTIVES IN GILT MANAGEMENT

Basically, what we require of the gilt is that she starts breeding activity (first heat at which gilt shows standing reflex and allows service by a boar) at a fairly early and predictable stage, that she will farrow down a good-sized litter at 10 to 11 months of age and a fairly moderate liveweight, that she will nurse her piglets well, that her body condition at weaning will not be too reduced so as to ensure prompt conception after weaning and she will proceed from there to thrive and produce good litters up to about the sixth litter.

While all these objectives cannot be fully achieved because of the incompleteness of our knowledge of the gilt, sufficient information is available to allow most pig-keepers to achieve more effective control of the gilt and her output than is the case at present.

REPRODUCTIVE DEVELOPMENT OF THE GILT

From its infantile stage at birth, the reproductive tract develops under the influence of hormones, mainly from the anterior pituitary gland.

The first milestone in reproductive development is reached when the gilt attains puberty, i.e. when she ovulates for the first time (see Plate 20). Ovulation is normally accompanied by oestrus or heat.

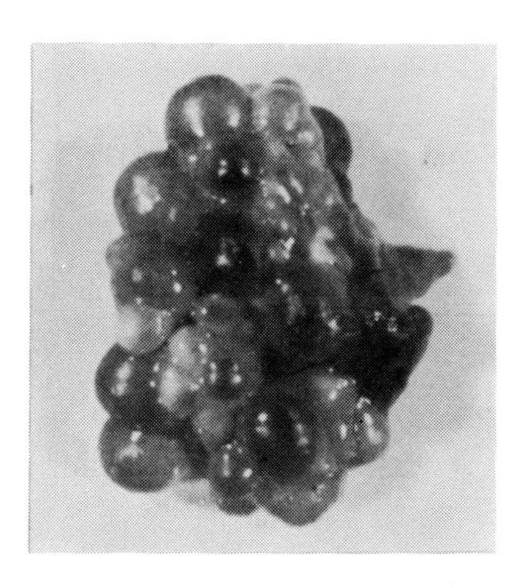

Plate No. **20**

Ovary of a gilt which has just ovulated for the first time. Note the many ova at different stages of development projecting from the surface of the ovary. The six largest projections are the positions from which ova have just been shed, the place of the ovum being taken by the *corpus luteum*.

Plate No. **21**

It is only in the presence of the boar that oestrus can be detected with certainty in all gilts.

Signs characterising oestrus or heat:

- Swelling and reddening of the vulva (usually but not always noticeable).
- The pricking of the ears in prick-eared breeds such as the Large White.
- By the gilt showing interest in the boar and the standing reflex, i.e. adoption of an immobile posture when pressure is applied to the back. In the absence of a boar, less than 50 per cent of gilts in heat will show the standing reflex. Only in the presence of, and under the stimuli of the boar, can heat be detected in the remainder.
- By the gilt allowing herself to be mounted by the boar or by other gilts. When the gilt is not in heat, she will usually squeal and run away either when the stockman applies pressure to the back or when the boar attempts to mount.
- By a characteristic grunt or 'roar' associated with heat.

It must be noted that swelling and reddening of the vulva by themselves cannot be taken to indicate a true oestrus accompanied by ovulation. Often such reddening and swelling of the vulva can be seen to recur two to three times in maiden gilts at approximately 21-day intervals before the true ovulatory heat or puberty appears. Thus, while there are many signs indicating oestrus, it is only in the presence of a boar that oestrus can be detected with certainty in all gilts. Once puberty is attained, oestrus will recur at 21-day (range 18–24 days) intervals until the gilt is mated and conceives.

FACTORS INFLUENCING AGE AT PUBERTY

Gilts are normally reared in the finishing pens along with pigs destined for slaughter, either for the pork, cutter, bacon or heavy pig markets, and replacement gilts are normally finally selected at the stage their contemporaries are slaughtered.

From observations made in pig units producing bacon pigs (approximately 90 kg liveweight at slaughter), while many gilts show recurring swelling and reddening of the vulva at approximately 21-day intervals in the finishing pens, very few gilts are capable of breeding before slaughter at bacon weight.

In a study at the Aberdeen School of Agriculture, a large number of gilts from nine different breeding/finishing units producing pigs for the bacon market had their reproductive tracts examined at slaughter. As indicated in Table 5·1, only 12 out of 1,201 gilts or 1 per cent had, in fact, ovulated and were therefore capable of fertile breeding before slaughter. The average liveweight at slaughter of this group was 89 kg and the average age 182 days with a range from 130 to over 240 days.

TABLE 5·1. Details of breeding activity in gilts slaughtered at bacon weight

Number of gilts examined	1201
Number which had ovulated once	12
Percentage which had ovulated once	1·0%
Average age at slaughter (days)	182
Average liveweight at slaughter (kg)	89

The gilts in this study had been kept under a wide variety of management systems and were representative of various purebreds, crosses and commercial hybrids.

It seems obvious that there is unlikely to be any marked breeding activity in gilts housed in conventional finishing pens up to bacon weight (approximately 90 kg liveweight). A feature of the findings in the above study was that very little difference was noticeable between herds in the degree of breeding activity up to bacon weight. This was surprising because it was known that there were quite marked differences between some of the herds examined in the age at which breeding activity starts. The difference was as much as 50 days in age at puberty between two of the herds, and this indicated that the treatment of the gilt after selection at bacon weight can have a big influence on the early onset of breeding activity.

Several factors are known to influence age at puberty which a producer wishing to stimulate earlier breeding activity, can exploit.

Among these factors are:

- Genotype.
- Nutrition during rearing.
- Stress associated with transportation or transfer and mixing of strange gilts.
- Contact with the male.

Other factors such as the lighting pattern to which gilts are subjected in rearing, season of the year, climatic conditions and group size are suspected of having an influence on age at puberty but their exact influence has not yet been established.

The role of the four factors listed above which are known to have an influence on age at puberty will now be briefly discussed.

Genotype

Breed differences in age at puberty are known to exist but, from the practical viewpoint, the most relevant facts are that inbreeding tends to delay attainment of puberty, while crossbreeding is associated with earlier puberty. It appears likely that, under the same conditions and management, the first or second cross gilt will attain puberty some 20 days before the mean of the constituent purebreds.

Nutrition during rearing

Regarding the effects of food or energy intake on age at puberty, it appears that normal levels of food restriction over the 55 to 90 kg liveweight range for bacon pigs is unlikely to delay puberty by more than one week relative to *ad libitum* feeding over this period.

While much work remains to be done on the effects of specific nutrients, in particular amino acids, during rearing on age at puberty, it is unlikely that nutritional factors will have a marked influence on age at puberty, unless they are correcting an obvious deficiency.

So, if gilts are being reared on a well-balanced diet and a feeding regime suitable for the finishing of their contemporaries destined for slaughter, it is unlikely that such a feeding regime can be greatly improved in order to induce earlier puberty.

Stress associated with transportation and mixing of strange gilts

It is a well-known phenomenon among practical pig-keepers that four to seven days following transportation or transfer and regrouping of gilts in strange accommodation around the expected age of puberty (about 170 to 220 days), a high proportion of gilts come into heat. The mechanism involved following such moving and mixing of gilts is thought to be the influence of associated stress affecting the output from the adrenal and pituitary glands of those hormones which stimulate reproductive development, culminating in ovulation and heat.

Contact with the male

Timely introduction of the male has been shown to bring forward the breeding season of sheep and to hasten puberty in rodents. Work largely at the University of Nottingham demonstrated that timely introduction of a boar to a group of maiden gilts was also effective in hastening puberty. At 165 days of age, Large White x Landrace first-cross gilts were regrouped, and while one group was given access to a boar, the other group was given no such contact. The results obtained are illustrated in Figure 5·1.

It can be seen that in the group having contact with a boar, 11 out of the 12 gilts reached puberty within 10 days of mixing and boar contact while the remaining gilt reached puberty at 183 days of age. There was no such dramatic response in the 'no boar influence' group and only 7 out of the 12 gilts reached puberty by 220 days of age. Thus, the influence of the boar appeared to hasten puberty, on average, by some 40 days.

The influence which the boar has in hastening puberty in this way is through chemical substances or pheromones produced in the submaxillary or preputial gland. The most likely pheromone exerting this influence is androstenol which is concentrated in the submaxillary gland. The boar, in champing its jaws in the presence of the gilt, probably releases this pheromone. The main influence on the gilt is therefore that of smell, while the sound of the boar also seems to be important. It would also appear that direct contact between the boar and gilts is essential for maximum stimulus; for when gilts have been penned along with a boar, the effect in stimulating and synchronising puberty has been much stronger than in an equivalent group of gilts in an adjacent pen which are being subject to all stimuli including sight, sound and smell of the boar and contact through a wire-mesh pen division. While the gilts in the latter group are being affected by strong boar stimuli, they are, of course, not actually penned along with the boar; consequently, the intimate interactions involving the boar nosing and making mounting attempts on gilts in the same pen would appear to be essential for full stimulation.

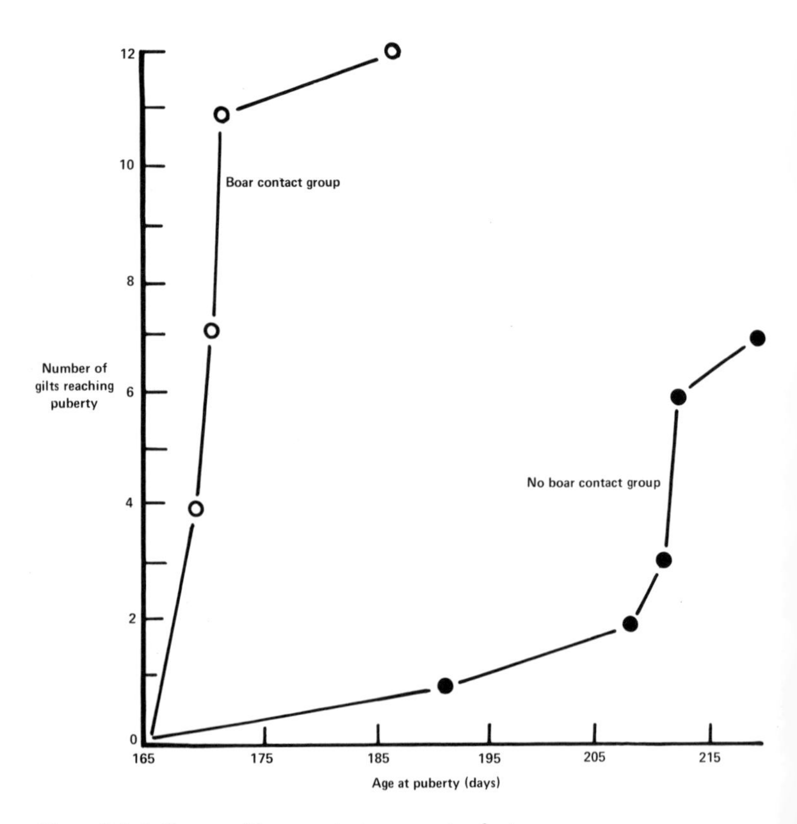

Figure 5·1. Influence of boar contact on age at puberty.

Source: Brooks, P. H. and Cole, D. J. A. (1970), University of Nottingham.

Producers have also found that when gilts around the expected age of breeding are transferred to a pen adjacent to that of a working boar in which many services are taking place, there is no dramatic response in such gilts in terms of stimulation of puberty, indicating again that such indirect contact of boars and gilts often fails to produce the desired response in gilts.

It was thought, mainly on the basis of experience with other species, that in order to maximise the effect of the boar in stimulating earlier puberty in gilts, exposure should not take place at too early a stage. The fear was that 'habituation' could occur, i.e. the prepubertal gilts would become accustomed to the presence of the boar and, as a result, the boar would have less influence in stimulating reproductive activity. However, recent work at the University of Nottingham indicates that gilts being reared to bacon weight (90 kg liveweight) along with contemporary castrates and boars do not experience 'habituation'.

Under most circumstances in the UK it would appear that the optimum stage to introduce a boar to crossbred gilts (Large White and Landrace) is around 165 to 170 days of age and some 15 to 20 days later for purebreds.

If, after introduction of a boar to gilts, they fail to start breeding activity, the boar may be failing for one reason or another to provide the necessary stimuli and should be replaced.

Older, more mature boars, probably because of their stronger 'boar smell', are much more effective in stimulating gilts than younger boars. Of course, after stimulation by an older boar, the latter can be replaced with a younger boar for mating. A young boar should be allowed access to maiden gilts under supervision as otherwise he can be attacked to such an extent that it can have an adverse effect thereafter on his confidence and ability to breed.

INDUCING EARLIER PUBERTY IN PRACTICE

If a producer wishes to induce earlier puberty in gilts, all the stimuli known to be effective should be combined in an attempt to produce the desired response.

As already indicated, Large White x Landrace gilts should be stimulated first around 165 days of age while the purebreds should not be stimulated until some 20 days later. This is because purebreds take longer to reach sexual maturity.

Taking the importance of proper timing into account, producers aiming for earlier puberty should therefore expose gilts of the appropriate age to the following stimuli:

- Transportation or transfer to new accommodation.
- Mixing with strange gilts.
- Contact with mature boar.

If boar contact involves the boar being in a pen adjacent to gilts, the combination of the above stimuli is likely to be partially successful in inducing earlier puberty, but the degree of success in obtaining earlier and better synchronised puberty can be improved by having direct contact between the gilts and the boar. Such a direct contact can be arranged by allowing a mature boar access to gilts for, say, half an hour each day from the start of the stimulation period until a response has been obtained. To avoid gilts being served at first heat and/or prevent services being unrecorded, supervision of boars during this contact period is essential and this increases labour costs. An alterntive is to house a vasectomised boar in the same pen as the gilts. What must be kept in mind is that vasectomised boars can vary in their stimulating properties and some display little interest in gilts. They are also fairly expensive to maintain and have to be replaced fairly regularly as they can become too heavy and injure young gilts; thus their usefulness is strictly limited.

FACTORS AFFECTING LITTER SIZE

Among the factors known to influence litter size from gilts are:

- Age and weight at mating.
- Breed or cross.
- Flushing.
- Timing of mating.

Other factors known to affect litter size include disease and nutritional deficiencies. Hormone therapy has also been used in attempts to improve litter size by increasing ovulation rate and reducing embryonic mortality but findings are not yet sufficiently promising to warrant practical application.

Some quite rightly contend that hormone therapy should be actively discouraged as it may result in the retention of sub-fertile females, their true capabilities being masked by therapy.

The role of the four factors listed above, which are known to have an influence on litter size, will now be briefly discussed.

Age and weight at mating

Over a fairly wide range, neither age nor liveweight by themselves appear to have a major effect on litter size of the gilt. This is demonstrated in Table 5·2 in which data is presented from different countries.

As is shown, data from some countries indicate no increase of litter size with increasing age at conception. From other countries (Ukraine and Britain), there is only a slight increase in numbers born alive per litter with increase in age at conception.

Data from a large breeding unit indicates that for purebred Large White and Landrace gilts, for each four weeks' delay in conception between the ages of 170 and 270 days, numbers born alive per litter increased by just under half a pig per litter.

TABLE 5·2. Relationship of age and weight at mating to litter size of gilts

Age at conception (days)			*Number born alive per litter*			*Source of data*
(1)	*(2)*	*(3)*	*(1)*	*(2)*	*(3)*	
210	240	270	9·0	8·6	8·8	Canada
181–200	201–230	231–260	9·8	8·9	9·3	Norway
181–210	211–240	241–270	8·1	8·1	9·0	Ukraine*
190 and under	191–220	221–250	9·3	9·6	9·4	Sweden
186–215	216–245	246–275	9·55	9·75	9·89	Britain

* Average liveweight at mating in this work was 65·5, 79·3 and 99·3 kg respectively for the three age ranges specified. Relevant liveweights for the other sources were not available.

As the gilt gets older, she will obviously be heavier at conception and farrowing, and the incentive to delay mating beyond a certain minimum age or liveweight is not great in view of the fairly small increase in litter size likely to be achieved.

The actual number of pigs born alive in relation to age and/or weight at mating probably masks earlier events; for, as the gilt increases in age and liveweight, there appears to be a fairly substantial increase in ovulation rate but this is accompanied by increased embryonic losses, with the result that the increase in litter size is considerably smaller than the increase in ovulation rate.

Litter size at first farrowing is more influenced by the heat number at which the gilt is mated than by her age or liveweight at that stage. Number of eggs shed is lowest at first heat and tends to increase by about one ovum from first to second and by a further ovum from second to third heat. Because of higher embryonic losses with increasing age and weight, the advantage in terms of number of pigs born as one delays service to the second or third heat is much less than the advantage in terms of ovulation rate.

While findings vary, on average an extra 0·4 piglet can be expected by delaying service until second heat, and the same increase in numbers born is likely by serving at third rather than second heat.

Since conception rate tends to be lower following service at first heat, service at second heat is to be preferred. Less of an argument can be made for delaying service from second to third heat.

Breed or cross

Breed differences in numbers born are well established. However, it is well known that numbers born per litter is influenced more by husbandry factors than by breed. The most marked genetic influence on numbers born relates to the advantage of the crossbred over the purebred sow, the crossbred producing about 5 per cent more pigs at birth (about half a pig per litter), on average, than the mean of the constituent purebreds.

Flushing

If gilts are to be mated at second heat, the most effective treatment is to avoid excessive food intake in the latter part of rearing, to increase their feeding level 10 to 14 days before mating and to reduce feed allowance to normal restricted levels immediately after mating.

The reason for such a feeding system is that both *ad libitum* feeding during rearing and high-level feeding after mating increase embryonic losses. Raising feed level by 50 or even 100 per cent from normal restricted amounts for about 10 days before mating has been shown to increase ovulation rate by up to two ova and this has resulted in an increase of up to one piglet born in some trials.

This process of increasing feed or energy intake for several days before mating is known as flushing. It is most effective in increasing numbers born to gilts when it follows a period of restricted feeding. The effect of flushing applied in this way varies between herds, but the response to flushing is likely to be greatest in herds where average litter size born to gilts is low, i.e. in the situation where an increase is most required and welcome. One of the effects of flushing is likely to be in ensuring gilts in heat receive adequate food just before ovulation. Their feeding drive is lower when in heat and reduced food intake because of excessive competition on group feeding when food is limited can adversely affect ovulation. Flushing helps to prevent this problem. The investment in feeding an extra 10 to 20 kg of food in the ten days or so prior to mating is not high and is a cheap insurance which can often produce a worthwhile return.

Timing of mating or insemination

Insemination or mating very early or very late in heat results in lower conception rate and litter size. The optimum timing of insemination to maximise conception rate and litter size appears to be 10 to 20 hours before ovulation. However, the exact timing of ovulation cannot be predicted, and so mating or insemination cannot be timed to produce an optimum result. However, when several inseminations or matings are arranged in the heat period, there is more chance that one of these will be optimally timed than where just a single mating or insemination occurs. Thus, double or triple service spread over the period when the gilt shows the standing reaction to the boar or to back pressure from the stockperson helps to increase litter size. This aspect is discussed in more detail in Chapter 10.

A SUGGESTED GILT MANAGEMENT SYSTEM

The factors influencing age at puberty, synchronisation of puberty in a group of gilts and litter size have been discussed above. We are now in a position to exploit this knowledge to achieve greater control over the gilt, to increase her output and to increase the efficiency of the breeding herd in general.

A suggested management programme from purchase (or selection) to service for a group of gilts is depicted in Figure 5·2.

This indicates that gilts would be best purchased or transferred from the bacon pens around 170 days of age, mixed with other gilts and given contact with a mature boar. Contact must be direct by allowing a mature, entire boar into the pen for at least 15 minutes each day under supervision. A better response will be obtained in some situations by leaving the boar for 24 hours per day with the gilts for the first four to five days until the first gilts show signs of coming into oestrus. At this stage, the boar should be removed to prevent uncontrolled services at first heat and placed in an adjacent pen from which he can be introduced to the gilts for a short period each day under supervision.

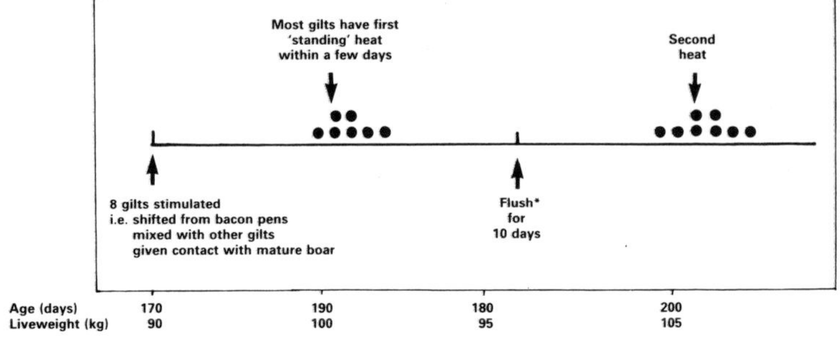

Figure 5·2. Outline of a management programme for gilts which has worked well in practice on some units.
● *Indicates one gilt on heat.* * *Flushing involves doubling the feed level for about 10 days before service or feeding to appetite.*

Crossbred gilts may be stimulated somewhat earlier than 170 days, but the optimum time for stimulation of purebreds is likely to be around 180 days.

Some five to ten days after such stimulation most gilts in the group can be expected to have their first 'standing' heat. This response is likely to be synchronised in that most gilts in the group will come on heat within about seven days of each other. In the unlikely event of the desired response not occurring within ten days, then the contact boar should be changed.

The above practices to stimulate earlier puberty and to synchronise first heat in a group of selected gilts work very well in some units but not so effectively in others. The precise reasons for these differences between units are not yet well understood. However, despite this variation in response from unit to unit, the above practices are the best natural ones to apply to achieve better control of gilt breeding.

That gilts can be predictably stimulated in this way provides the necessary control required in most situations over their breeding activity to facilitate a positive policy regarding the culling of defective sows and the programming of replacement gilts into the breeding herd. The synchronisation of the response of the gilt to the stimuli imposed means that second and third heats will also follow in a synchronised manner. This facilitates flushing, i.e. increasing the level of feeding some ten days prior to the heat number at which the gilts are to be served (usually second or third). This practice has been demonstrated to increase ovulation rate and number born in some herds, in particular those with low average numbers born to gilts.

Implementation of 'flushing' in practice is extremely difficult if grouped gilts reach puberty over a protracted period but is greatly facilitated by synchronising first heat as described above. It must be remembered that, as soon as gilts are served, high-level feeding should be discontinued so as to minimise embryonic losses. Unless all gilts in a group are served at the same time, individual feeding facilities will be required so that the food intake of newly-served gilts can be controlled.

Objections to such a system of gilt management as depicted above may arise on two grounds:

1. That an increase in litter size of the gilt is likely to reduce her body condition to such an extent during lactation, that difficulties are likely to be experienced in getting her to rebreed and conceive quickly after weaning.
2. That the gilt bred at a younger age and lower liveweight may have a poorer lifetime performance.

These are indeed very real and justifiable fears but the anticipated problems need not arise and these aspects are dealt with in the following section on 'Rebreeding of the Gilt'.

Before proceeding to this section, however, an alternative approach to synchronsing oestrus in cycling gilts (gilts which have had at least one 'standing' heat) will be outlined. This involves the use of the progestagen allyl trenbolone (Regumate, Hoechst UK Ltd). This product should be included in the ration of gilts at a rate of 20 mg daily over an 18-day period. Such treatment has the effect of synchronising the oestrus cycles of the gilts within a group and the great majority of gilts have their next heat within two to three days of each other, some eight days after the withdrawal of Regumate. Thus, this is likely to become a useful technique in practice for synchronising heat in a group of gilts which had already been cycling prior to the start of treatment. It should be noted that the treatment is not designed for gilts which have not yet had their first heat.

REBREEDING OF THE GILT

It is recognised that greater problems are experienced in getting sows to conceive after weaning the first litter than after subsequent ones.

It is natural that excessive loss of weight and body condition during the first lactation should be blamed for this problem; this is indeed likely to be a contributory factor since the young gilt is still maturing rapidly and there is more competition for nutrients as between growth, lactation and reproduction than in the mature sow.

Therefore, there is a need to try to minimise such loss of body condition in the gilt when she is suckling her first litter. One approach is to content oneself with only a small first litter, of say seven to eight. However, this is a defeatist attitude and there is a need to look for ways of increasing food intake in the first lactation.

Overfeeding gilts in pregnancy reduces appetite in lactation, so that if gilts are too fat at their first farrowing, one cannot expect the gilt to eat as much in lactation as if she was fit but not fat at farrowing. Another way of increasing intake in the first lactation is to employ an *ad libitum* feeding system by fitting a small hopper on to the front of the farrowing crate. This will result in higher intakes relative to the conventional system of feeding restricted quantities twice daily, and the target weight change during lactation can be achieved by controlling the length of time the gilt is *ad lib* fed. Those who find that high levels of feeding around farrowing are related to post-farrowing fever complex leading to agalactia (milk shortage), must obviously avoid feeding *ad lib* in the first few days after farrowing. Feeding wet rather than dry and providing a higher energy diet are other ways of increasing intake in lactation.

Work was undertaken at the University of Nottingham to examine the usefulness of increasing the feed level after weaning for gilts which had lost a lot of condition in the first lactation. Results are presented in Table 5·3.

TABLE 5·3. Effects of increasing feed levels after weaning the first litter

	Daily feed levels from weaning to mating (kg)		
	1·8	2·7	3·6
Percentage conceiving within 42 days of weaning	58	75	100
Weaning to first heat (days)	22	12	9

Source: Brooks, P. H. and Cole, D. J. A. (1975), University of Nottingham.

The usual feeding level after weaning was 1·8 kg and increasing this produced a useful response in this situation. However, it is desirable to try to prevent excessive weight loss in lactation by the steps outlined above. Irish workers have shown that the level of crude protein and lysine in the lactation diet is also critical in relation to the rebreeding interval after the gilt is weaned (see Chapter 11).

Trying to economise by feeding a diet with less than 14 per cent crude protein and 0·6 per cent lysine can often be false economy during lactation.

LIFETIME PERFORMANCE OF THE GILT

Does the gilt bred at an early stage have lower lifetime performance?

Data from one large commercial herd indicated that purebred gilts conceiving before 190 days of age were more likely to be culled before their third litter than gilts conceiving later. This may, however, have had less to do with age at conception and more to do with the increased likelihood of such gilts losing a lot of condition in their first lactation culminating in rebreeding problems. Avoiding such excessive loss of body condition by taking the steps suggested in the previous section may help to reduce this possible problem.

It was established earlier that by imposing the appropriate stimuli on maiden gilts at the right time one can induce first true heat some 40 days earlier than in gilts treated in a conventional manner. This provides a producer with a potential advantage in that, whether his policy is to serve at second or third heat, he can get his gilts served some 40 days earlier and at a proportionately lower liveweight. This reduces the length of the unproductive rearing period and, in theory, it should result in a relatively smaller sow throughout its productive life; this in turn should result in lower maintenance requirements, and, if output can be maintained following such early breeding, then overall efficiency can be increased relative to breeding at the conventional stage.

Some of the concrete answers to much of the above speculation have been provided by results of work at Seale Hayne Agricultural College in which gilts mated early (approximately 190 days) and those mated some 40 days later (controls) at conventional age and weight have been compared. All gilts were mated at second heat. Results are presented in Table 5·4.

Efficiency of food utilisation was improved by over six per cent in early-mated animals. Thus, from this preliminary work, a system based on earlier mating of gilts is showing up well in terms of sow output and efficiency of food use relative to a conventional system. However, for those producers who, for one reason or another prefer to mate their gilts at a more conventional age and liveweight, stimulation of earlier puberty using the methods described is still likely to be very worthwhile. For if gilts have their first heat at a younger age and lower liveweight they will be more mature in a reproductive sense by the time they are served at the conventional stage and benefits are likely to accrue in terms of improved conception rate and larger litter size.

SUMMARY

The gilt warrants much more attention than she receives at present in the average sow herd. By using certain stimuli (transfer to new premises, regrouping and boar contact), at the appropriate age, gilts can be stimulated to have their first heat some 40 days earlier than usual and at a proportionately lower liveweight. Not only is earlier breeding activity achieved but a group of gilts so stimulated will tend to have their first heat within a few days of each other. The ability to predict timing of first heat and the synchronising of this heat makes possible a positive approach to gilt management in that knowing when gilts will have their first, second and third heats allows a producer to programme gilts into the herd and makes possible a positive culling policy. Synchronisation of first heat also makes it easier to flush a group of gilts for about ten days prior to mating whether this is to be at second or third heat. In herds where litter size of gilts has been poor, flushing can increase numbers born.

TABLE 5·4. Performance over five parities of sows on two systems of mating

	Early mating	*Conventional age/weight at mating*
Total pigs born	53·7	53·8
Pigs born alive	51·6	50·4
Pigs weaned	42·6	43·8
Litter birth wt (kg)	64·0	60·1
Litter weaning wt (kg)	388·2	399·1
Total sow plus piglet food (kg)	2349	2508
Mean piglet birth weight (kg)	1·20	1·13
Mean piglet weaning weight	9·16	9·13
Sow food per piglet weaned (kg)	55·7	58·3
Sow food per kg weaner (kg)	6·1	6·5

Source: Brooks, P. H. (1977), Seale Hayne Agricultural College, Devon.

Gilts which have been stimulated to reach puberty earlier may be served at first, second or third heat according to need. Mating at first heat is likely to be associated with smaller litter size and poorer conception rate. There will be a slight increase in litter size by delaying mating from second to third heat but, in general, such a delay is unlikely to be worthwhile. Provided stimulated gilts mated at second heat are managed adequately thereafter, their lifetime performance and efficiency will not be adversely affected and may well be improved.

As a result of somewhat scant attention in the past, the gilt has not been as productive a member of the sow herd as she might have been. There have been difficulties in predicting when she will breed and this has not encouraged a positive culling policy so that either many suspect sows have been kept for one litter too many or else this has resulted in expensive sow accommodation not being fully used. Many gilts have been excessively old and heavy at first mating and both these factors reduce the efficiency of sow output. Litter size born to, and weaned by, many gilts has been relatively low and, with gilt litters constituting 20 to 30 per cent of total litters in most herds, this has resulted in often quite serious depressions in output and efficiency from the herd as a whole.

Many of these inefficiencies associated with the gilt can be overcome by exploiting the increased knowledge on how to effect greater control over her breeding activity and performance as outlined in this chapter.

Chapter 6

FARROWING

PARTURITION, OR FARROWING, is one of the most critical stages in the whole process of pig production, for the well-being of both the sow and the piglets. Various problems can arise which can result in death or, at least, reduced efficiency of both sow and piglets. It is therefore important to recognise what constitutes normal parturition so that departures from normal can be detected quickly and prompt remedial action taken.

GESTATION PERIOD

Gestation period of the sow (counting day of first service as day nought) is normally given as 114 days, or, to aid memories of students in particular, three months (one month = 30 days), three weeks and three days. The mean gestation period in different herds can vary between 113 and 116 days depending on genotype and management. In any given herd, gestation period can vary over a fairly wide range as indicated in Figure 6.1, which is based on a total of 1,542 consecutive farrowings in a large commercial herd.

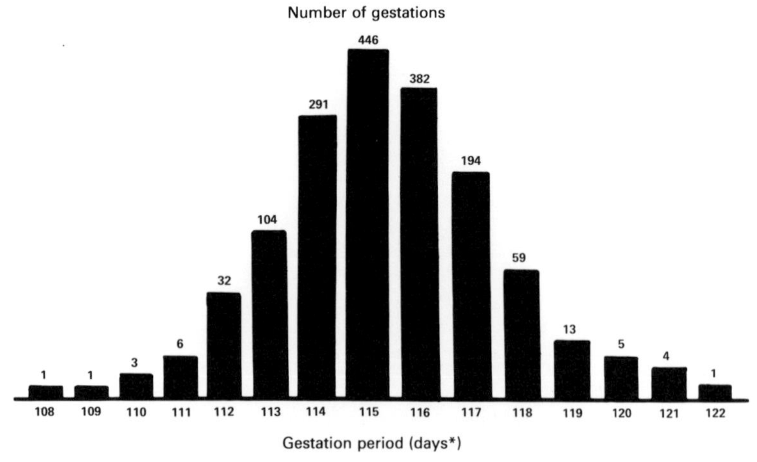

Figure 6·1. Frequency distribution of gestation period.

Thus, while gestation period in this herd had a mean of just over 115 days and had a very wide range from 108 to 122 days, 92 per cent of farrowings occurred within two days of the mean, i.e. from 113 to 117 days. 98·9 per cent of all farrowings occurred within four days of the mean, i.e. from 111 to 119 days, leaving only 1·1 per cent farrowing outside this period.

The range of gestation period makes it essential that sows are transferred to farrowing quarters well before the due-to-farrow date which is based on the herd mean (115 days in this case). If the standard advice had been followed to the effect that sows should be transferred to farrowing quarters one week before the due-to-farrow date, all sows in this herd would have been transferred in time.

The precise factors causing variation in gestation period are not known. Differences between herds in mean gestation period do exist and these are probably caused by differences in both genotype and management. Within a herd, there seems to be a fairly close relationship between numbers born per litter and gestation period. This is shown in Figure 6·2.

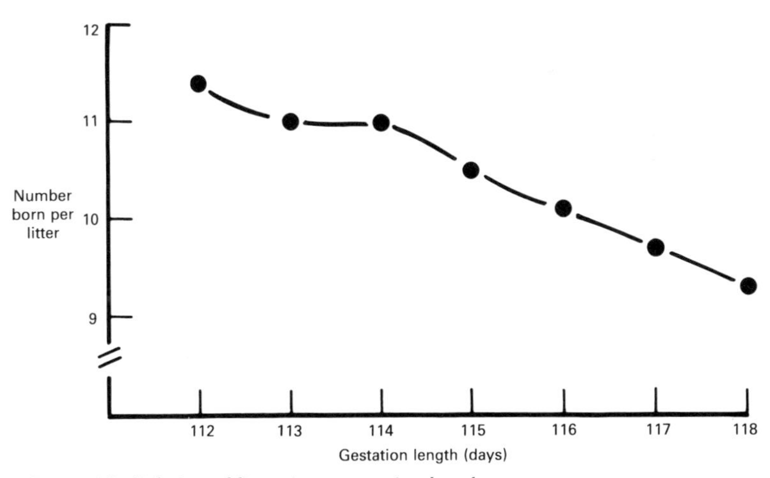

Figure 6·2. Relation of litter size to gestation length.

Source: Pig Improvement Company Ltd.

This data was also derived from a commercial farm and shows that the smaller litters tend to have a longer gestation period. The factors leading to larger litters being farrowed earlier may be stimulated by the greater weight of uterine contents, or, more likely, by the greater hormone production of the larger litter.

PHASES OF FARROWING

Farrowing can be divided into three phases:
Stage 1. The preparatory stage (dilation of the cervix).
Stage 2. The stage during which the piglets are expelled.
Stage 3. The stage during which the placenta or afterbirth is expelled.

First stage
The cervix (muscular entrance to womb) dilates in preparation for delivery of the piglets and the muscular walls of the womb start to contract rhythmically, moving the foetuses towards the pelvic inlet. These contractions occur about every 15 minutes and last about 5 to 10 seconds but become more frequent as farrowing progresses. During the first stage, the only outward signs are those of abdominal discomfort and restlessness.

Second stage
As soon as the first piglet enters the cervix, which lies within the bony canal or pelvis, the sow begins to assist expulsion of the foetus with visible abdominal contractions (straining or pressing). Our observations indicate that the period from the beginning of abdominal contractions to expulsion of the first piglet may take from one to three hours in most cases (with a range of 15 minutes to almost 10 hours).

Third stage
The ends of each afterbirth become fused with the adjacent afterbirth in the majority of cases, forming a continuous tube through which the piglets pass. However, part of the 'tube' may break away during farrowing so that a clump of afterbirth may be passed. Nevertheless, in the majority of cases, all the afterbirth is passed in one mass after the last piglet has been born.

SIGNS OF IMMINENT FARROWING

It is important to be able to predict the onset of the second stage of farrowing (i.e. expulsion of piglets) accurately in situations where supervision of farrowing is a routine practice. There are several indicators that farrowing is imminent and these include:
- Abdominal contractions (as discussed above).
- Bed-making.
- Increased restlessness.
- Expulsion of blood or blood-stained fluid from the vulva.

- Expulsion of meconium from the vulva.
- Twitching of the tail.
- Change in rectal temperature.
- Texture of the udder and availability of milk.

The usefulness of the above criteria in helping to predict timing of onset of the second stage of farrowing (expulsion of piglets) will now be discussed.

Abdominal contractions

As discussed above, these appear in most cases about one to three hours prior to the birth of the first pig, the observed range being 15 minutes to 10 hours prior. With such a wide range, this factor cannot predict the birth of the first pig with any degree of accuracy.

Bed-making

As farrowing approached, the wild sow busied herself gathering together dry herbage, selected a well sheltered spot and made a nest in preparation for farrowing and for the young litter. This instinct is still present in the domestic sow when she is provided with bedding. When no bedding is provided, the sow tends to paw the floor of the farrowing pen as if going through the motions of making a bed.

Our observations indicate that bed-making can start any time from one hour up to 22 hours prior to the birth of the first pig, the mean being about five hours. Others have observed bed-making in sows up to three days prior to the birth of the first pig. After the sow starts bed-making by collecting litter from all parts of the pen within reach, she will arrange and rearrange this bedding at fairly regular intervals up to the birth of the first pig.

Because of the wide range in timing relative to the expulsion of the first pig, bed-making is not a good predictor of onset of farrowing.

Increased restlessness

From being a very contented, peaceful animal in late pregnancy, the sow reaches a crescendo of restlessness as farrowing approaches. There is an increase in activity as a result of bed-making, the sow may chew any available structure in the pen, probably as a result of discomfort, nervousness and frustration, while she drinks, urinates and defecates more frequently. Her noise level also increases as farrowing approaches.

Expulsion of blood-stained fluids

About 40 per cent of sows show no obvious signs of blood-stained fluid at the vulva prior to the birth of the first pig. Blood-stained fluids can be observed being expelled from the vulva in the remaining 60 per cent of sows, an average of some 100 minutes before the first birth. The observed range in expulsion of blood-stained fluid was from 15 minutes to six hours before the first pig with about 90 per cent of sows showing such a discharge when within two hours of farrowing.

Thus, while some sows show no discharge of blood-stained fluid prior to the birth of the first pig, when this discharge is observed, the chances are very high that the first piglet will be born within two hours from such an observation.

Expulsion of meconium

The small, greenish-brown pellets expelled in the fluids, or surrounding piglets at farrowing, are meconium or foetal faeces. In about a quarter of farrowings, some meconium is expelled prior to the birth of the first pig. In such cases, meconium can be seen being discharged from the vulva an average of 40 minutes (range 1 to 100 minutes) before the first birth. Thus, when meconium is seen, there is unlikely to be long to wait before the first piglet is expelled.

Twitching of the tail

Tail twitching is a phenomenon displayed by sows which would appear to be connected with expulsion of piglets and afterbirth from the genital tract to the exterior. As the sow has a contraction, so the tail is pulled right back through an angle of about 180° out of the way of the vulva and this is often accompanied by brisk to-and-fro lateral movements.

Tail twitching was observed an average of two hours prior to birth of the first pig, the range being one minute to 10 hours before. But the great majority of cases of tail twitching occurred within two hours of the first birth.

Change in rectal temperature

As farrowing approaches, respiration and pulse rate are gradually accelerated, sow activity increases and body temperature gradually rises by about 0·5°C (about 1°F), on average, in the 10 hours prior to the birth of the first pig. However, the temperature rise is not great or consistent enough in the case of an individual sow for this criterion to be a useful predictor of time of delivery of the first pig.

Texture of udder and availability of milk

As farrowing approaches, the udder changes from a soft and

flabby condition to being fairly firm and turgid. Also, as farrowing approaches, the sow is more likely to respond to light and gentle massage of the udder. Such gentle rubbing of the udder, especially the anterior glands, can stimulate the sow to lie down if standing and to rotate her trunk so as to expose all her teats as she lies on her side, issuing the contented grunts characteristic of nursing at the same time. When in this contented frame of mind, the udder may be tested for presence of milk. If the sow is angry and unco-operative when her udder is massaged, one cannot effectively assess the presence of milk with a view to predicting onset of farrowing, because in this situation the output of the hormone adrenalin is likely to be inhibiting output of the hormone oxytocin which is responsible for 'let down' of milk.

When the udder of the co-operative sow is massaged and an attempt made to express milk from the teats (preferably the front teats), various stages of milk availability have been noted. Sometimes as farrowing approaches, on squeezing a teat, a thin almost clear fluid may be observed either in pinhead-sized drops or in large blobs. If milk can be expressed, it may appear as pinhead-sized drops, large blobs or may come squirting from the teats. Milk in pinhead-sized drops and large blobs has been observed in some sows up to five and three days prior to farrowing respectively. Milk can be squirted from the teats of some sows up to 24 hours before farrowing but, in most cases, milk cannot be expressed in this form until some eight hours prior to the birth of the first pig.

Most sow behavioural aspects which suggest that parturition is imminent, such as bed-making and abdominal contractions, are not evident continuously from the time that each one is first displayed, whereas others, such as availability of milk, are detectable at any time. Signs in the latter category are more useful predictors of farrowing in cases where the onset is checked by regular short visits to the farrowing quarters since, during these, the signs which occur only periodically may not be displayed.

PREDICTION OF ONSET OF FARROWING FROM GENERAL OBSERVATIONS

The foregoing factors are not the only ones which may be used in an attempt to predict the timing of birth of the first pig. As farrowing approaches, the vulva swells, the pelvic ligaments relax and the udder drops while the pulse and respiration rates increase.

The factors which may be used to predict onset of farrowing have so far been discussed individually. In practice, of course, prediction of onset of farrowing is not made on the basis of these factors considered individually but of them all considered together. And while it has been observed that several of the factors considered can be useful individual predictors of timing of birth of the first pig, there is little doubt that the onset of farrowing can be much more accurately forecast by the simultaneous consideration of all impending signs.

BEHAVIOUR OF THE SOW DURING FARROWING

The marked increase in the average activity of sows as they approach farrowing, continues during the second phase, i.e. when piglets are being expelled. There are exceptions to this rule in that some sows are very peaceful in the second phase and lie on their sides throughout. However, other sows can be very restless and this particulary applies to gilts. This restlessness, of course, places newborn piglets at risk from overlying, especially since the piglets themselves are in the process of unsteadily and somewhat randomly exploring their new environment, with a tendency to keep in fairly close contact with the sow. If the sow is standing, newborn piglets are quite likely to be groping about underneath her where they are at obvious risk from crushing should she lie down. Alternatively, they may be suckling or asleep at the udder where they are at risk should she roll over from lying on one side to the other.

After Phase Two of the parturition is complete and the last piglet has been born, the sow becomes decidedly more peaceful and settles down to nurse her piglets. The difference in activity between stage two of farrowing when the sow is expelling her piglets and subsequently is shown in Table 6·1.

TABLE 6·1. Activity of 31 sows in different stages of farrowing

	Stage 2 (first to last pig)	*Stage 3 (expulsion of afterbirth)*
Number major movements* per hour	2·0	0·34
Number minor movements† per hour	3·1	0·7
Proportion of time spent lying on side	84	95

* Major movement = lying down from dog sitting or standing positions.
† Minor movement = turning from side to side, from belly to side or from side to belly.

It can be seen that, in terms of major and minor movements, sows were six and four times respectively more restless in Stage

Two than in Stage Three. The increased peacefulness after the end of Stage Two is also indicated by the fact that sows spent a considerably higher proportion of their time lying on their side in the suckling position in the third than in the second stage of farrowing.

It is interesting that in this early period the risk of piglet death or injury from overlying was almost as great from minor as from major movements of the sow.

There is a tendency on the part of sows to be much more restless at the start of the second stage of farrowing than towards the end of this stage. Sows have a habit of standing up or adopting a dog sitting position after each of the early piglets are born as if to examine the new arrivals.

It may be this apparently exploratory examination of earlier born piglets which is sometimes exhibited in savaging.

Attempts at savaging are probably more common than supposed because farrowings are rarely attended throughout and therefore attempts at savaging can escape notice.

In a sample of 31 farrowings which were attended and observed throughout, but in which neither sows nor piglets were interfered with in any way, the incidence of savaging was as shown in Table 6·2.

Table 6·2. Incidence of savaging attempts* in 31 farrowings

	Litter number		
	1 (Gilt)	*2 to 4*	*6 to 9*
Number sows (farrowing)	9	12	10
Number making savaging attempts*	8	3	2
Percentage making savaging attempts	89	25	20

* Savaging attempt = an aggressive attempt by the sow to bite a piglet.

It was not the policy in this herd to cull sows because of savaging attempts so this does not explain the decreased incidence of savaging in older sows.

Of the eight gilts which made savaging attempts, four made only one savaging attempt, two made two attempts, one made eight attempts and one made 13 attempts. Only in the case of the gilt that made 13 attempts were any pigs gripped by the dam sufficiently to be injured, there being two piglets slightly injured in this particular case. Both were reared successfully and, in proportion to their birthweight, had reasonable weights at weaning. Of the three sows of intermediate age (second to fourth litters) involved in savaging, one second-litter sow made three savaging attempts and another second-litter sow made four attempts. A third-litter sow made five savaging attempts. No piglets were injured as a result of savaging attempts by these sows of intermediate age. One seventh-litter sow made one vain savaging attempt and an eigth-litter sow made four savaging attempts in one of which one piglet was slightly injured but it survived and had a reasonable weight at weaning. The latter sow savaged her first litter very badly and only one out of a litter of six survived. Token attempts at savaging were made by her in subsequent litters but no piglet in these litters died as a result of savaging.

With regard to the timing of savaging attempts, these were usually made immediately the first pig was born. In the case of many gilts and also of some older sows, one got the impression that this reaction was due to fear as the first-born piglet began to move about and issue small high-pitched grunts. The gilt or sow would jump up to her feet and attempt to turn round in her farrowing crate to try to attain the piglet, champing her jaws angrily in the process. When the piglet came within reach the dam might make a token snap at it. Often, once the dam had seen and smelled the piglet, she would settle down and make no more savaging attempts. Thus, savaging was directed against only the first-born piglets except in those few cases where many savaging attempts were made when most or all of the litter were often involved.

It may be that the intervention of an attendant in the case of savaging could worsen rather then improve the mood of the gilt. However, it is only natural for an attendant to intervene in a case of savaging and an injection of a suitable tranquilliser is usually all that is required to settle the gilt down onto her side so that the piglets already born can proceed to get a suckle in safety and the gilt can proceed with farrowing the remainder of the litter. Azoperone (Stresnil) has been found to be suitable for this purpose in most cases but where this does not produce the desired response, it may be necessary to muzzle the sow. There is some evidence that savaging is more common in certain families then in others.

DURATION OF FARROWING

The length of the second stage of farrowing varies markedly between sows. In the sample of 31 farrowings quoted the findings were as shown in Table 6·3.

TABLE 6·3. Duration of second stage of farrowing and interval between births

	Litter number			
	1 (Gilt)	*2 to 4*	*6 to 9*	*Overall*
Number farrowings	9	12	10	31
Period first to last piglet (mins)	86	180	141	140
Range first to last piglet	42 to 143	61 to 374	56 to 267	
Period between births (mins)	12	15	21	16
Range in mean period between births per litter	6 to 21	7 to 34	5 to 81	
Range in period between births for individual piglets (mins)	0 to 56	0 to 136	0 to 177	

Thus, in this small sample, the second stage of labour took an average of 140 minutes to complete and the average interval between births was 16 minutes. Interval between births was shortest for gilts (12 minutes) and longest for older sows (21 minutes). There was a considerable range in the overall duration of Stage Two of farrowing (from 42 to 374 minutes), in the average interval between births for the separate litters (6 to 81 minutes) and for individual pigs (0 to 177 minutes). Only two sows took longer than five hours to complete the second stage of farrowing, while only a handful of piglets had a birth interval exceeding 60 minutes. While farrowings in which the second stage lasts longer than five hours can result in birth of normal viable piglets, and while piglets born more than one hour following the previous birth can be perfectly viable, the figures of five hours for duration of Stage Two and 60 minutes for interval from previous birth should be regarded as upper limits. When Stage Two is showing signs of proceeding beyond five hours and when one hour has passed since the birth of the last piglet and the sow continues to strain, consideration should be given to intervening in the process as all may not be well. Abnormalities in farrowing and appropriate action to take are covered in a subsequent section.

The shorter interval between births in the case of gilts may indicate that they have better muscular tone than older sows and therefore abdominal contractions may be more effective and, as a result, they can expel piglets more effectively. Other studies carried out some 13 years ago at the same time as the above study reported a similar average interval between births of about 16 minutes. More recent work indicates an average interval between births of some 25 minutes and it is possible that this could reflect a change from loose housing to individual stalls for dry sows. The decreased opportunity for exercise in dry sow stalls might result in inferior muscular tone, making the piglet expulsion process less efficient. However, this is sheer speculation, although, taking the human analogy, childbirth appears to be a much more straightforward and natural process in the underdeveloped countries where women are more involved in activities which help to maintain good muscle tone than is the case in the relatively pampered and sedentary Western societies.

PIGLET BEHAVIOUR AND BIRTH 'PHENOMENA'

Following birth, piglets get to their feet within minutes and instinctively make attempts to reach the udder and to suckle. On average, piglets obtain their first successful suckle in about 45 minutes following birth.

Piglets may have different birth presentations, some may be born with their umbilical cords ruptured, some may be born still completely enveloped in the foetal membranes and the order in which they are born is also of practical importance. The influence of these various factors on the newborn piglets is indicated below.

BIRTH PRESENTATION

As distinct from other domesticated species, the 'hind legs first' presentation is as normal and common as the 'nose first' presentation in piglets. In one study involving birth of 337 piglets in 31 farrowings, 52 per cent were hind feet first presentation, the remainder being nose first. Most other studies show that almost half of piglets are born hind feet first. Any departure from nose or hind feet first presentation is termed malpresentation and includes piglets being delivered broadside on and tail first only with the two hind feet still retained.

While in one study piglets coming hind feet first took about one minute more to be delivered, on average, following the previous birth, such piglets suckled as quickly and grew and survived as well as those born nose first.

STATE OF UMBILICAL CORD AT BIRTH

The functioning umbilical or navel cord is the lifeline from the mother to the foetus. It is the channel through which the foetus is supplied with nutrients and oxygen and which is responsible for elimination of waste products from the foetus.

The umbilical cord is elastic in nature and is capable of considerable stretching. Thus, most piglets are born with the cord still attached to the placenta and some may even reach the udder with the cord still intact. When the cord is attached at birth, it takes about four minutes, on average, before the cord is broken and the piglet is freed from this attachment. In about 20 per cent of live births, the cord is already broken at the moment of birth. Of course, the cord may have broken immediately before birth, or it may have severed up to about five minutes previously. Piglets born towards the end of farrowing are more likely to have their cords severed at farrowing than those born nearer the start. Although piglets with severed cords at birth take slightly longer to get their first suckle, they appear to survive and grow as well as those born with intact cords.

INTERVAL BETWEEN BIRTHS

It has already been stated that the average length of the second stage of farrowing is some 140 minutes and the average interval between births about 16 minutes. However, there is a considerable range in these values and, while a piglet may be born immediately after the previous birth, other piglets may be delivered as long as three hours or even more after the previous one. The longer the apparent delivery period of a piglet, the longer it takes to free itself from the cord attaching it to the placenta and the longer it takes to get a successful suckle. Thus, a long delivery period would seem to have an adverse effect on the vigour of the piglet at birth.

While there is no hard and fast rule, if up to one hour has passed since the farrowing of the previous pig and the sow is showing signs that she still has piglets to deliver, then the stockman should consider exploring the situation and giving assistance as necessary, exercising every possible precaution so as to avoid injury or introducing infection. Prolonged birth interval can result in piglets which were alive at the start of farrowing, dying from suffocation in the uterus. Such deaths are termed intrapartum stillbirths or Type 2 stillbirths as distinct from Type 1 stillbirths or prepartum deaths, which have occurred before the start of farrowing.

INTERVAL SINCE FIRST PIG

All piglets in a litter may be delivered in about half an hour or may take five hours or more. The longer the period from time of delivery of the first pig, the longer piglets take to reach the udder and to suckle. This delay in suckling after birth with increase in delivery time from first pig may indicate that such a delay has depressed the vigour of the newborn piglet, but this is also likely to be partly due to the greater physical difficulty later-born piglets have of getting to the udder and to obtain a successful suckle because of the presence at the udder, and competition from, earlier-born piglets.

Interval from birth of the first pig is related to the birth order and this will now be discussed.

BIRTH ORDER

The incidence of stillbirths which occur during the farrowing process (intrapartum deaths) increases with the birth order. A typical situation is depicted in Plate 22 which shows a litter of 16 arranged in the order of birth. Three intrapartum deaths occurred in this litter and these came 11th, 13th and last in the litter.

These piglets dying from anoxia (suffocation) intrapartum can often be recognised by the staining of their body with greenish-brown material which is meconium or foetal faeces. These faeces are released as the piglet suffocates in the uterus and they soil the surface of the piglet.

The very much higher incidence of intrapartum deaths towards the end of farrowing is illustrated in Table 6·4.

TABLE 6·4. Incidence of intrapartum deaths according to birth order

Birth order	*First half*	*Middle pig*	*Second half*	*Total*	*Third last pig*	*Second last pig*	*Last piglet*	*Per cent of all intrapartum deaths in last 3 births*
Number intrapartum deaths	4	1	39	44	8	6	17	70·5

Thus, out of 44 intrapartum deaths, 39 occurred in the second half of farrowing and 31 (or 70·5 per cent) were among the last three piglets born.

Later-born piglets therefore run a greater risk of being born dead as a result of happenings during farrowing. The reason for this is primarily the greater risk of piglets being suffocated in the uterus as farrowing proceeds.

There are several explanations for this. It has already been mentioned that later-born piglets are more likely to be born with their umbilical cords (or lifelines) broken than their earlier-born litter-mates. If the cord broke only shortly before delivery, the piglet may be none the worse. However, if the cord broke some time before delivery, then obviously the risk of the piglet being

The penalty for being last in the farrowing queue!

Numbers between photographs indicate the time interval between births in minutes

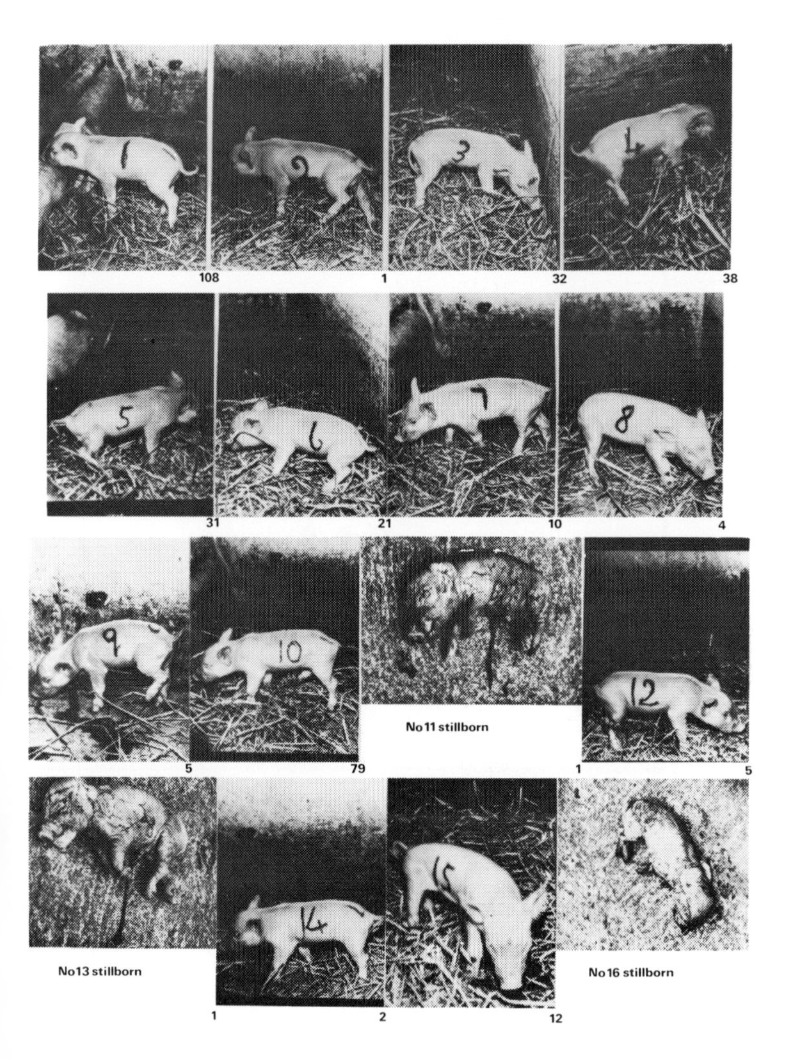

Plate No. **22**

A litter of 16 arranged in order of birth. The increased incidence of stillbirths towards the end of the litter is typical.

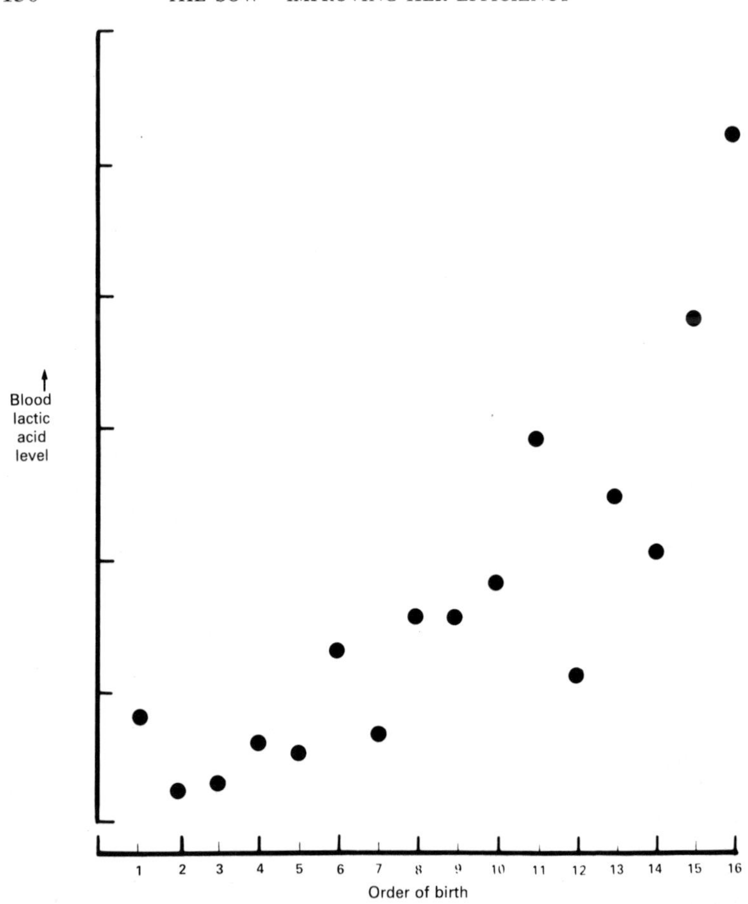

Figure 6·3. Degree of anoxia (suffocation) in liveborn piglets at birth according to birth order.

dead through suffocation by the time of delivery is much increased.

It must also be kept in mind that the uterine contractions which start before the first pig is delivered continue and even intensify as piglets are being delivered. These contractions serve to push the piglets, the foetal fluids and the placenta or afterbirth towards and through the birth canal to the exterior. Each contraction, by exerting pressure on the umbilical cords, may reduce the blood flow to the piglets, thus reducing the amount of oxygen reaching the unborn piglet. This causes partial suffocation and, with each successive contraction, the effect is likely to worsen until a stage may be reached, if delivery is unduly prolonged, where the piglet dies from suffocation, thus contributing to the incidence of intrapartum deaths or type two stillbirths. Another factor which may contribute to this problem as farrowing proceeds is the increased likelihood of the placenta becoming detached from the uterus prematurely before all piglets are delivered. This also effectively cuts off oxygen supply from mother to foetus and results in death through suffocation of the unborn piglets. Accidents such as knotting, twisting or haematoma of the cord have similar effects.

It is clear, therefore, that farrowing can be an inefficient process and result in deaths of unborn piglets through suffocation. Such intrapartum deaths are about four times as numerous as prepartum deaths (dying before farrowing starts) and can approach an average of almost half a pig per litter or one pig in every second litter. This can amount to a loss of some 200 pigs in a 200-sow herd in a year and, on a national scale in the UK, almost one million pigs per year are lost through suffocation during the farrowing process.

Losses through intrapartum deaths, therefore, should give cause for concern and the fact that these pigs are dying so soon before delivery increases the challenge to decrease this important source of loss.

This must be done by increasing the efficiency of farrowing and, in particular, trying to avoid unduly long intervals between births and length of farrowing in general. With this in mind, the use of drugs such as neostigmine to help stimulate more effective uterine contractions during farrowing are proving promising in research work and the possible usefulness of this technique is discussed in more detail later in this chapter.

The events during farrowing which increase the likelihood of later-born piglets being suffocated in the uterus and therefore being born dead, also appear to decrease the survival chances of later-born piglets. It can be shown that later-born livebirths have been deprived of oxygen during delivery to a greater extent than earlier births. This is shown in Figure 6·3.

The amount of lactic acid present in the blood of the piglet at birth gives a useful measure of anoxia or the extent to which the piglet was deprived of oxygen (or suffocated) during delivery. It can be seen that later-born piglets have been deprived of oxygen during delivery to a greater extent than earlier-born piglets.

Later-born piglets are also at a disadvantage since their earlier-born litter-mates will have had greater opportunities for suckling colostrum which provides advantages in terms of both disease resistance and nutrition. Moreover, later-born litter-mates are faced with much greater competition in getting into the udder and actually obtaining a suckle than earlier births.

For these various reasons, earlier-born piglets have greater chances of survival than their later-born litter-mates. It is not easy in practice to do much to improve the lot of later-born piglets. The more efficient the process of farrowing, the better will their survival chances be. Farrowing can take much longer in older sows so that timely culling of such sows can help in this respect. In order to give later-born piglets equal chances of suckling and of obtaining adequate colostrum, it has been suggested that piglets should be placed in a comfortable creep as born and deprived of access to the sow until all piglets have been farrowed, at which stage they can all be given access to the udder in an attempt to give equal chances for obtaining colostrum. However, the suckling of earlier-born piglets induces output of the hormone oxytocin and this same hormone helps to stimulate contractions of the uterine muscle and therefore aids in prompt delivery of piglets. So by keeping earlier-born piglets away from the sow, it is conceivable that the process of farrowing could be prolonged.

Thus, there are indications that the sow might not perform the job of farrowing all that efficiently and her inefficiences will be reflected in intrapartum stillbirths and perhaps also in liveborn piglets whose viability may have been adversely affected by partial suffocation during delivery.

PIGLETS BORN ENVELOPED IN AFTERBIRTH

Some piglets are born alive partially or completely enveloped in their placenta or afterbirth. In a study on 31 farrowings involving 326 liveborn piglets, five such piglets were born while two others were partially enveloped, in one case the hind part of the body being covered and in the other the fore part was covered. Thus, the incidence of this occurrence was about two per cent. Of the seven piglets, five were the last-born in their litter, while, of the remaining two, one was born fifth in a litter of seven and the other was tenth in a litter of 16. Thus, there is an obvious tendency for these 'enveloped' or 'partially enveloped' piglets to be born near the end of the birth order.

Because of the study involved, these piglets were not interfered with in any way but left entirely to their own devices. Six out of the seven succeeded in escaping by virtue of vigorous struggling. The two partially enveloped piglets escaped in less than 30 seconds after birth. Of the five totally enveloped piglets, four succeeded in breaking through their covering and escaping after 10, 15, 30 and 180 seconds. The unfortunate exception struggled vigorously but vainly for 11 minutes after which there was no further sign of life. Except for one of the escapees which died from atresia ani (blind gut), the remainder survived and grew normally to weaning.

It may well be that the problem of piglets being born alive in the afterbirth is frequently overstated since the incidence in this small study was only two per cent and the great majority of enveloped piglets (six out of seven) were able to escape when left to their own devices.

SOW AND PIGLET BEHAVIOUR – SUMMARY

Piglets have a very strong instinct to reach the udder, those born with intact umbilical cords free themselves from this attachment some four minutes after birth, on average and achieve their first successful suckle some 45 minutes after birth.

Some sows are extremely restless during the second stage of farrowing, i.e. the period between the birth of first and last piglets and this, combined with the piglets' determined and somewhat random attempts to reach the udder, or stay as close to the sow as possible if she is standing, renders them very prone to injury or death from overlying.

Piglets have almost an equal chance of being born nose or hind feet first. Abnormal presentations include breech first with hind legs retained or piglets coming broadside on. Earlier-born piglets are less likely to have their cords ruptured at birth than their later-born litter-mates.

Time taken to expel all piglets averages some 140 minutes but may take up to five hours. Some concern should be felt when farrowings proceed beyond five hours. Average interval between births is about 16 minutes and when up to 30 to 40 minutes have passed from the birth of the previous pig and the sow shows signs of not having farrowed all piglets, then again, there is some cause for concern.

Later-born piglets are at a disadvantage relative to earlier-born litter-mates. They are more liable to die in the uterus from suffocation during farrowing (intrapartum stillbirths) and later-born piglets which are born alive are slightly less viable, on average, than earlier-born litter-mates. This is caused partly by the increased degree of suffocation suffered during delivery by later-born piglets and possibly also by the greater difficulty they have of obtaining a suckle and of obtaining adequate colostrum.

Some sows savage their piglets and this characteristic is most evident in gilts. The basic cause of this may be fear and nervousness.

DETECTING PROBLEMS DURING FARROWING

A very low proportion of sows are assisted at farrowing and this is probably due to the small size of the foetus in relation to the birth canal. However, because such a high proportion of stillbirths are of the intrapartum type, i.e. dying during farrowing, more careful monitoring of farrowings followed up by assistance as required, would be likely to pay useful dividends on many pig units. A variety of factors can lead to problems at farrowng.

As farrowing approaches, illness of the sow, signs of farrowing without the onset of birth and a foul, offensive vaginal discharge are all singular indications that veterinary advice should be sought immediately.

After the birth process has begun, if up to 30 to 40 minutes elapses from the previous birth and the sow continues to press and strain, physical examination is desirable.

Prompt action is required when there are indications that something is interfering with the normal process of birth.

Failure to expel a foetus by the dam is called dystocia. Dystocia may be maternal in nature (fault of the sow) or foetal in nature (fault of the piglet).

FAULTS IN THE SOW

The uterus (womb)

When the sow looks as if she should be farrowing but does not enter the second stage of labour, the case may be a primary failure of the womb to contract (primary inertia). There are many suggested causes for this, for example calcium deficiency, disease of the womb, over-distention due to many foetuses and hormonal imbalance, to name but a few.

Should this problem be suspected, seek veterinary advice.

Occasionally the womb will cease contracting due to exhaustion (secondary inertia), for example, near the end of a large litter or after a prolonged attempt at expelling a large piglet. Such a problem is more common in old sows because of the loss of muscular tone in uterine muscles. In the case of some older sows towards the end of farrowing, often piglets can collect in a deviation of the womb just behind the pelvis but it may be some considerable time before their delivery is completed. This delay puts such piglets at risk of dying during delivery and ending up as intrapartum deaths. Regular monitoring of the birth process followed by provision of timely and appropriate assistance as required is called for in such cases. If piglets cannot be removed manually, veterinary advice should be sought.

A drug which gives promise of assisting the safe delivery of later born piglets in big litters, particularly in older sows, is neostigmine. This drug stimulates contraction of smooth muscle including that in the uterus and this could be useful in assisting delivery of later-born piglets. In work in the University of Aberdeen, when 5 mg Prostigmin (Neostigmine methylsulphate, Roche Products Ltd) has been administered to sows following the delivery of the fourth or fifth pig, piglet delivery time following injection has been decreased by about nine per cent while stillbirth rate has been reduced by 20 per cent. The only side effect which was noticeable was slight vomiting in less than 20 per cent of the sows some 30 minutes following injection.

On fairly rare occasions, serious problems can arise during farrowing.

Sometimes one horn of the womb may twist completely round at its junction with the other horn, so preventing the expulsion of its contents. This can only be diagnosed by physical examination of the birth canal. You should therefore seek veterinary advice. Do not interfere if this condition is suspected.

The birth canal

Faults may be either in the bony tissue surrounding the soft canal or in the soft tissues of the canal. The bony canal may be reduced in diameter because of inherited or developmental factors or simply due to a previous fracture which has healed. Examples of problems in the soft tissues are tears in the vagina, haematoma and distended bladder.

If birth cannot be assisted manually, veterinary assistance should be sought.

Hysteria

Perhaps due to fear or excitement, this condition is confined to gilts and results in total inhibition of the farrowing process. If this occurs, administer sedative or seek veterinary advice.

FAULTS IN THE PIGLET

Presentation

Piglets are normally born either nose or hind feet first and any departure from this is termed malpresentation. Two piglets coming together, a piglet coming broad side on, siamese twins, tail first only with the two hind feet still retained, and two-headed piglets, are examples.

If any such faults occur, it is advisable to seek veterinary advice if the piglet cannot be physically removed.

Foetal oversize

This occurs when the piglet is too large in relation to a birth canal of normal size. This can happen in the case of immature gilts.

You can attempt manual removal carefully but if this fails you should seek veterinary advice.

WHEN PHYSICAL INTERFERENCE IS INDICATED

Any competent pig producer should be capable of assessing the situation, making an internal examination and assisting the removal of piglets.

Assessing the situation

Firstly, the pig-keeper should be aware of the normal train of events. Secondly, he or she must be able to assess his or her own capabilities in relation to the situation. Knowledge and experience are the two vital factors here. Once a problem has been recognised, it is important that a decision should be taken as soon as possible. Prevarication and delaying tactics never lead to success. When a wrong decision has been taken it is better that the pig person should swallow his or her pride and seek veterinary advice immediately.

Making an internal examination

Suitable protective clothing should be worn and the hand and arm thoroughly washed in disinfectant and warm water. The hand and arm should then be lubricated with a mild soap or suitable obstetrical preparation. The vulva and surrounding area should be washed clean. Ensuring that fingernails are trimmed, the fingers of the hand to be used should be introduced to the passage in a cone shape (this is easier with the sow lying on her side).

After making sure that the arm cannot be trapped by an obstacle such as a bar on the back of a farrowing crate, the examination proper should begin. The hand is gradually passed along the passage palpating the wall of the vagina, then the cervix, which is continuous with the vagina, until either a foetus is felt or the hand can no longer penetrate the womb. Should resistance be felt, it is advisable to withdraw the hand and lubricate thoroughly before trying again. At no time should brute force be used. The bony circle of the pelvis can easily be felt but in the majority of cases the piglet will be found beyond the rim of the pelvis. Difficulty may be experienced while passing a hand through the bony portion of the canal in gilts. Once the piglet has been palpated, its situation in relation to the canal should be ascertained.

Assisting removal

If no gross abnormalities of the canal or piglet can be detected and minor malpresentations have been corrected, the piglet should be firmly gripped and gradually withdrawn. Both hind legs should be grasped firmly or the head may be held behind the ears or just within the bony rim of the eye sockets. Great difficulty may be experienced withdrawing both the head and hand through the bony pelvis at the same time. In such cases a stiff but malleable loop of wire (electric flex, for example) may be placed behind the head of the piglet. Very little effort should be required to remove the foetus, no more than is necessary to lift a 3 kg weight.

After the successful removal of the piglet, another exploratory examination should be made. Sometimes, several piglets can be removed in quick succession. The afterbirth (placenta) of one piglet is fused at its end to the adjacent afterbirth of the next piglet and hence several afterbirths tend to be passed in a clump or all may be passed after the last piglet has been born. Veterinary assistance should immediately be sought if the piglet cannot be removed manually by the pig person. Should this decision be delayed, the sow's life may be endangered as well as that of the foetuses.

AFTER-CARE OF PIGLET

Many foetuses requiring manual removal will be anoxic (partially suffocated) and some will already have inhaled mucus into the wind pipe (trachea). *Do not* give the kiss of life to begin with; this will only force the sticky mucus further down the wind pipe into the lungs.

Action

- Firstly, the nose and mouth should be cleared of mucus, a soft rubber tube being useful here.
- Swing the piglet violently round at arms' length, making sure that no objects are within range!
- Then give the kiss of life, if necessary.
- If a piglet's heart is beating at birth but it is showing little sign of breathing, immersing the body up to the neck in a container of cold water can often stimulate the pig to gasp, and thus normal breathing can be initiated.
- The piglet should be dried off quickly and placed under or over the source of heat.

AFTER-CARE OF SOW

When farrowing has been manually assisted, it is wise to administer an injection of antibiotic. Occasionally your vet will prescribe a longer course of treatment in cases of internal injury or when piglets have been dead and infected for some time. The sow should be eating and drinking within 12 to 15 hours of farrowing, but occasionally normal gilts will not eat for up to 24 hours after farrowing. The udder should be checked for agalactia (lack of milk) and mastitis (inflammation of the udder) for at least three days running. The rectal temperature should be checked on the day following the farrowing. The normal temperature of the sow is 38·9°C (102°F) and while it is normal for temperature to rise to 39–40°C (103–104°F) just after farrowing, a rise over 41°C (105°F) should be viewed with concern and appropriate treatment administered promptly. This treatment should be that which has been found by pig unit staff and their vet to be most effective in dealing with similar problems in the past.

INHERENT VIABILITY OF NEWBORN PIGLETS

The viability of a piglet at birth is a function of its state before farrowing began and of any adverse effect such as injury or partial suffocation imposed on it by the farrowing process.

The viability of the piglet before farrowing began will depend on its genetic make-up; presence of any genetic defects and disease; its weight, which will affect its ability to conserve heat when born (a smaller piglet has a large surface area in relation to its weight from which it can lose vital heat energy to its surroundings); and its energy reserves which will be related to its weight.

This inherent viability of the piglet before farrowing began can be reduced by adverse effects occurring during parturition which could result in death during farrowing (intrapartum stillbirths) as a result of injury, or, more likely, suffocation. If born alive, the piglet's chances of survival can be adversely affected by partial suffocation or injury suffered during farrowing.

Thus, newborn piglets will vary in their viability at birth according to inherent factors and those induced by problems occurring during farrowing.

The good stockman will have done his utmost to ensure development of viable piglets and will have taken all reasonable care, supported by his vet, to ensure as safe delivery as is within his power, at farrowing.

Once the piglets are born alive, the challenge to the stockman is:

- to be aware of the basic needs of these newborn piglets;
- to be aware of the dangers to which they are subject;
- to recognise that their viability at birth varies;
- to be aware of how to take this variability in viability into account in setting out to maximise piglet survival and, finally,
- to provide the newborn piglets with adequate conditions so as to cater for their basic and individual needs.

REQUIREMENTS OF THE NEWBORN PIGLET

The basic requirements of the newborn piglet are as follows:

1. Minimum challenge from infection and provision of resistance against infection prevailing in the herd.
2. Safety from such dangers as savaging and overlying.
3. An adequate environment.
   Newborn piglets have very limited energy reserves and these must be fully conserved by providing a very comfortable environment with adequate temperature and freedom from draughts. In this way, the maximum amount of energy will be available to enable the piglet to obtain regular and adequate nutrition and disease resistance (via colostrum) from the sow from immediately after birth; this will contribute to its optimum growth and health.
4. Adequate and regular nutrition.

One must set out to provide these simple but basic requirements by arranging a reasonable amount of supervision during farrowing with problems being detected quickly and remedial action taken as appropriate.

Providing adequate maternity/nursery facilities in helping to meet requirements 2 and 3 above is the subject of the next chapter. Some of the factors conducive to ensuring availability of adequate and regular nutrition are to be covered in Chapter 8.

It should be clear that in order to maximise survival of all viable piglets born, action is necessary on a variety of fronts. These various actions require to be incorporated into appropriate systems suitable for practical application. Accordingly, all components related to increasing piglet survival are brought together into proposals for a farrowing and rearing management system in Chapter 9.

Chapter 7

FARROWING QUARTERS—THE BASIC REQUIREMENTS

Newly born piglets are very vulnerable. They have low energy reserves and can die fairly quickly as a result of chilling unless provided with an adequate temperature and freedom from draughts. They are extremely small in relation to the size of the dam and are therefore very prone to death from overlying by the sow. The sow can be very restless and awkward during the process of farrowing and this further endangers the newly-born piglet.

FARROWING QUARTERS – BACKGROUND

The need for specialised care and attention for the sow and piglets at, and for some time after, farrowing has long been recognised. Specialised facilities for crating or tethering the sow and providing comfortable safe creeps for piglets have been provided for decades in a variety of ways. Despite this, it is true to say that we are little nearer a standard optimum design now than was the case 40 years ago.

PLANNING FARROWING ACCOMMODATION – CONSIDERATIONS

The major factors to be considered in planning farrowing accommodation are as follows:

- Welfare of the sow and piglets, including cleanliness.
- Ease of observation and supervision.
- Labour economy (e.g., forward creeps to improve piglet accessibility for handling and slatted floors to achieve self cleaning).
- Capital investment.

Some of these considerations are antagonistic to each other; for instance, some types of slatted and/or perforated floors keep sows and *surviving* piglets clean and healthy but they can result in a high incidence of deaths from crushing soon after birth and can also result in foot and leg lesions among very young piglets. It is impossible in a single arrangement within a pen to cater fully for all objectives, so that a compromise in design is necessary. No two people will put the same emphasis on the same factors. Some will place most emphasis on low capital investment possibly at the expense of piglet welfare and well-being, some may go for maximum labour economy, while others will put maximum weighting on sow and piglet welfare at the expense of higher capital investment and labour costs. The outcome is an infinite variety of systems in practice, all having their strong and weak points.

GENERAL SITUATION

From the above, it is clear that there is no optimum system to meet all requirements. A point which must be made, however, is that in planning farrowing accommodation in the recent past, there has been too much emphasis on labour economy at the expense of the adequate welfare of sow and piglets. We all recognise the high cost of present-day labour but over scrimping on labour is often a case of 'being penny wise and pound foolish'.

ORIGINAL OBJECTIVES

The original objectives in providing specialist farrowing facilities are in danger of being lost sight of under present-day pressures to minimise capital investment and effect maximum labour economy. It would be wise to recall these original objectives and give them their due emphasis in planning modern farrowing facilities.

These objectives were:

1. To reduce incidence of crushing by:
   (a) controlling the movements of the sow within the pen;
   (b) attracting piglets to a comfortable safety zone or piglet creep;
   (c) catering for the different temperature requirements of sow and piglets by providing a higher temperature in the piglet creep.

2. To ensure adequate nursing and suckling, and therefore piglet nutrition by:
   (a) ensuring that the bottom bars of the farrowing crate were high enough and the crate wide enough to ensure that these bars imposed no impediment to piglets suckling;

(b) providing a comfortable floor surface to encourage the sow to rotate her body at nursing so as to expose fully all teats on the lower row to her piglets;

(c) arranging an adequate microenvironment for piglets in the creep area.

These original objectives of minimum piglet losses and adequate thriving of piglets must receive high priority when arranging modern farrowing facilities although they must be achieved in such a way as to meet, as far as is possible, present-day requirements for labour economy, ease of management and minimum capital investment.

OBJECTIVES AND THEIR COMPONENTS

1. *Minimising deaths from crushing*

The important components in this objective are as follows:

- Contented sows.
- Contented and thriving piglets.
- Care on the part of the sow in changing her posture.
- Adequate design of the farrowing crate to help the sow control her posture changes more effectively, especially in lying down.
- Good inherent mobility in piglets and provision of non-abrasive floor materials which provide no impediment to free movement of piglets.
- Provision of sufficient attraction in the creep area/areas (in the form of heat, light and comfort) to ensure that piglets are attracted out of the danger area within the area of the farrowing crate except when suckling.

2. *Adequate nursing and suckling, and therefore, nutrition*

The important components in this objective are as follows:

- Healthy sows and piglets.
- Height of the bottom bar above floor level in relation to providing full and adequate access of piglets to the top teats.
- A comfortable floor surface to encourage the sow to rotate her body at nursing so as to fully expose all teats on the lower row to her piglets.
- The width of the crate in relation to the size of the sow. This obviously also influences the accessibility of teats to piglets.

ACHIEVING OBJECTIVES

In designing a farrowing facility, how do we set about meeting the requirements of the farrowing sow and her piglets in their first critical few days of life in such a way as to minimise risk of crushing and maximise opportunities for adequate and regular nutrition on the part of every viable piglet born?

The farrowing pen components which must be considered to provide the necessary conditions for the sow and her newly-born piglets are as follows:

- The farrowing crate.
- The floor surface.
- The piglets' creep.
- Watering arrangements.
- Pen size.
- Heating arrangements.

These will now be dealt with in turn.

THE FARROWING CRATE

This should help to control the movements of the sow in such a way that she is forced to lie down on her belly before rolling over on to either side. Most crushing of piglets stems from sows flopping directly on to either side from the standing position trapping unsuspecting piglets. Thus, crates should be so designed to prevent sows from flopping in this way. Figure 7·1 on page 147 illustrates contrasting types of design in this respect.

Sows in the design in Figure 7·1(a), with more latitude for movement, are likely to tramp more of their dung through the slats in slatted or partially slatted pens and keep their pens slightly cleaner, but it is this greater latitude for movement which is likely to increase incidence of crushing in such pens.

The critical height and lateral spacing of the horizontal bars of the crate illustrated in Figue 7·1(b) is indicated in Figure 7·1(c). Farrowing crates with approximately the same profile as that in Figure 7·1(c) will assist the sow in controlling her descent from the standing position and so help to minimise risk of crushing.

The risk of crushing can be reduced even further by fitting farrowing cradles to such a crate (see Plate No. 23). In lying down, the sow normally lowers her front end first and then carefully lowers her hind-quarters. Without farrowing cradles, when her hind-quarters reach a level just below the second horizontal bar on the crate (approximately 450 mm above floor level), some sows tend to lose control and their hind-quarters can 'flop' down suddenly during this last part of their descent. This sudden 'flop' can put at risk any piglets in this danger area at the time and newly-born piglets are particularly vulnerable as they wander somewhat aimlessly around the area adjacent to the site of birth, seeking contact with the sow and her udder.

The farrowing cradles depicted in Plate No. 23 are designed in such a way that they can be readily attached to, and detached from, the main framework of the farrowing crate by the stockman. The main bar of the cradle (which should lie approximately 300 mm above floor level so as not to impede access of piglets to the teats) is hinged and made of strong but light material so that it is manipulated very easily by the sow as she lies down, rolls over on to her side, stands up and lies down again.

The cradles should be left attached to the farrowing crate for two to three days after farrowing until the danger period of crushing has passed.

While the cradles can be readily and quickly detached from one crate and attached to another, some producers prefer to leave them permanently attached to the crate. If properly designed and fitted, they appear to cause no inconvenience whatsoever to the nursing sow and her litter but merely oblige the sow to lie down in a more responsible manner at all times so as to minimise the risk of overlying her piglets.

In farrowing crates fitted properly with such well-designed cradles, in pens which are also well designed for the farrowing and lactating sow and her litter, deaths from overlying have been reduced to about one pig or less out of every five farrowings (0·2 or less of a piglet per litter). The great majority of such deaths from crushing take place during the farrowing process with virtually no deaths occurring thereafter from this cause.

Farrowing crates may be fitted with adjustable fronts and rear gates. In pens with only a narrow slatted area at the rear of the pen, adjusting the front for small gilts and sows helps to ensure that as much as possible of the dung and urine deposited lands on the slatted area. Some crates have adjustable bottom bars which can be raised or lowered according to the size of the sow. Lowering the bar can prevent small sows getting trapped underneath, while the bar can be raised for large sows to allow adequate access of piglets to all the teats. Side opening crates are available and can be useful where space is limited for a front and/or rear access pass. While a front access passage is useful for facilitating feeding and handling of piglets, a rear access pass is more useful at farrowing, since it is easier for the stockman to intervene and assist in farrowing without disturbing the sow. Rear access passages should not be less than 750 mm wide.

THE FLOOR SURFACE

The floor surface in a farrowing pen should:
(i) cater for good mobility of piglets;
(ii) eliminate risk of injury to piglets and the udder;
(iii) be comfortable for the sow and provide her with a good foothold;
(iv) provide an acceptable level of cleanliness and be easy to clean;
(v) be durable (i.e., have a life of at least ten years);
(vi) be of low cost (materials plus labour cost for laying) per year of life.

In some types of slatted flooring the hole size is too big in relation to the size of the piglets' feet. From work done in the North of Scotland College, it would appear that gap width should not exceed 10 mm to prevent small piglets getting their feet stuck. Some types of slat have sharp edges and this can result in lacerations of piglets' feet leading to risk of secondary infection. Such floors can also result in lacerations to teats particularly when the udder is in a turgid state just before and after farrowing. Such teat damage can effectively reduce rearing capacity.

Thus, caution is advocated in choosing slatted floors for farrowing pens from the point of view of damage to piglets and the udder and in relation to mobility of the newborn piglet.

Other important criteria in the type of flooring selected are the foothold provided for the sow and easy cleaning properties. Some types of welded mesh and woven wire appear to match up to all of these criteria reasonably well.

One important factor on which little information is available concerns the relative willingness and efficiency with which sows rotate their body to expose the teats on the lower row during suckling. Failure to expose teats on the lower row, especially the more posterior ones, is a very common phenomenon as illustrated in Plate 29B in Chapter 8. This problem is greater in older sows with more pendulous udders but it is likely to be affected by the type of flooring. It is likely that, in the case of some of the less suitable types of slatted floor, such as some types of expanded metal, the sow may rotate her trunk only with some discomfort as she tries to expose her botton teats and this may deter her from doing so effectively. There is little doubt that sows are most comfortable on a solid floor bedded with straw, sawdust or wood shavings and are likely to be most effective in exposing teats on the lower row at suckling in such a situation. Thus, the more

Figure 7.1 (a)

Figure 7·1
The design in Figure 7·1(a) allows the sow to flop directly on to her side, thus increasing the risk of crushing of piglets. The design in Figure 7·1(b) obliges the sow to lie down on her belly first before rolling over on to either side. The arrangement of the horizontal bars of the crate is illustrated in Figure 7·1(c).

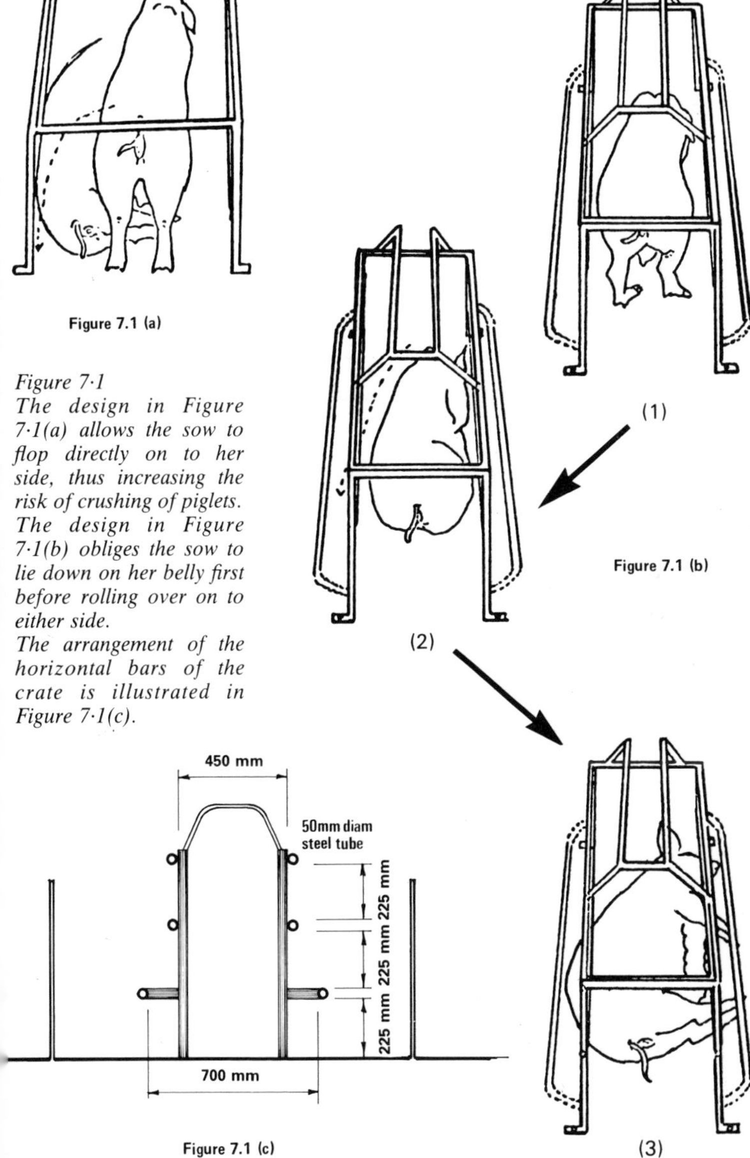

Figure 7.1 (b)

Figure 7.1 (c)

effectively types of slatted floors can approach the well-bedded solid floor in comfort, the more efficient that floor is likely to be for the mothering of the sow. Slatted floors of woven wire and welded metal have acquired a good reputation for providing the necessary degree of comfort, safety and cleanliness. More specific comments on alternative types of flooring for farrowing pens are detailed below.

Concrete

New floors should have a steel float finish. Having the correct concrete mix is vitally important (not too much or too litttle water) as is an adequate degree of compaction, to be finished with a steel float and then polished with the float to produce a smooth surface.

Curing should take place over seven days; the surface should be misted frequently with water and covered with straw, hessian or light polythene sheeting in dry weather. The surface should then be left for a further 14 days before exposure to livestock.

Further specifications

Materials	– Portland cement, fine natural sand, clean water	
Proportions	– By volume	1 cement:2·5 sand
	By weight	50 kg cement:150 kg sand
		(112 lb cement:336 lb sand)
Water to	– Dry sand – not to exceed 29 litres/50 kg (800 litres/cu metre)	
	Damp sand – approximately 22 litres/50 kg (640 litres/cu metre)	

Old floors which have become abrasive should either be rescreeded as for a new floor or treated with chlorinated rubber paint. This paint reduces the skid resistance, coefficient of static friction and abrasion. It lasts for about 10 weeks and is inexpensive. The aim in time is to find a more durable finish at approximately the same cost.

Flattened expanded metal

There are several grades of flattened expanded metal available which vary in relation to void size. Those with a void size greater than 10 mm in any one direction are not recommended for the area of the floor on which the piglets suckle, unless the void edges are protected with a coating of plastic or similar material. The plastic coating has the effect of reducing the void size and protecting the edge detail. However, plastic coating has not been found to be durable under the sow. Unfortunately, when the void size in expanded metal is less than 10 mm in any one direction, faeces tend to accumulate and have to be removed at regular intervals.

Plate No. **23**

Farrowing crate fitted with farrowing cradles to oblige the sow to lie down carefully on her belly before rolling over on to either side. The cradles are made of light but strong material. Being hinged, they are easily manipulated by the sow so that she experiences no difficulty in turning from side-to-side, in standing up and in lying down.

Punched metal

The sharp edge detail of punched metal does not meet the basic requirements and some types of punched metal have been dirty in practice.

Welded metal

From birth to three weeks of age, 12 × 75 mm mesh (5 SWG) is recommended. This has a void size of approximately 6 mm. Injury levels to piglets' feet and sows' teats are low on this type of flooring and cleaning properties are extremely good.

Woven wire

This material is very similar to welded metal but is probably slightly superior in terms of the grip afforded to the sow and piglets. Injury levels to both piglets' feet and the sow's udder are virtually non-existent. Cleaning properties are extremely good.

THE PIGLETS' CREEP

The creep should provide a temperature of 28 to 30°C (82 to 86°F) for piglets. Because of the high cost of heating and in order to minimise the effect of convection currents and draughts, the creep area should be well protected by providing an enclosed area which can be more easily heated while providing adequate access to the sow. Such an arrangement is illustrated in Plate No. 24, heating being provided either by an infra-red lamp or by a heat pad on the floor with a pilot light fitted inside the creep to attract the piglets at first.

It is also important that the heated creep be adjacent to the udder of the sow at the start since piglets have a strong instinct to lie as near the udder of the sow as possible in the early hours of life as shown in Plate 25.

In the first 48 hours of life, when piglets have heated side creeps, they use these areas for resting to a much greater extent than they use a heated front creep in a pen with such an arrangement. This is illustrated in Table 7·1.

TABLE 7·1. Proportion of total resting time spent lying in the heated creep area

Age of piglet (hours)	*Percentage using creep*	
	Pen with side creep	*Pen with front creep*
0–12	42%	14%
12–24	76%	23%
24–48	92%	47%

Source: Wilkinson and English (1980), University of Aberdeen.

This behaviour indicates that side rather than forward creeps are preferable for newly born piglets. Our observations indicate that

Plate No. **24**

A covered creep area bedded with wood shavings or good quality straw which is capable of providing a very high degree of comfort for young piglets. The box conserves the heat produced by the piglets and because such a small air space has to be heated, heating costs can be reduced appreciably. The lid of the box is hinged to allow easy inspection of the piglets.

in some types of farrowing pen with only a forward creep area, it can take at east 24 hours for some litters to start using the forward creep. In the early hours of life, they prefer to lie near the sow and, if there is no attraction to the side creeps adjacent to the udder, then they prefer to lie against the udder. In this position, they sleep within the danger area for crushing and the incidence of crushing is likely to be higher as a result.

If in these farrowing houses the side creeps are made more attractive to newborn piglets, e.g., temporary solid covering laid on top where an entirely slatted floor is used, with light and heat provided, then they will readily use the side creep on each side of the sow for the first 24 hours or so after farrowing, after which the side creeps can be made less attractive (e.g., remove light and heat to front creep) and the front creep made more attractive. Piglets are more likely to be attracted to a heated forward creep if it is lit by a heat lamp or a pilot light while the general lighting in the house is kept dimmed except when the herdsmen are carrying out their duties.

There is no doubt that piglets can be trained to use a front creep earlier by shutting them in this area from shortly after birth but this takes time and, despite this, the instinct to stay as close as possible to the sow and, more particularly the udder, during these critical first few hours of life still largely persists.

However, the attempt to train piglets to use a front creep area can be combined with precautions to reduce the risk of overlying. The lactating sow's most restless and excitable periods during the day are around feeding time and, this being so, it is a good practice for the first two to three feeds after farrowing to enclose the piglets in the safety of the creep area from the time the sow is fed until she has settled down afterwards. Thus, in pens with a forward creep, such a practice achieves the twin objectives of increasing the piglets' safety while at the same time encouraging them to use the heated creep area at an earlier stage.

WATERING ARRANGEMENTS

If nipple drinkers are used, they should not be mounted too high on the crate as otherwise, in drinking, the jowl and neck of the sow will act as a drainpipe as the sow drinks and any solid area in the front of the crated area will become wet. The nipple is best mounted just above the food trough so that any spillage ends up in the trough and not on the floor.

Water bowls are more costly than nipples but possess some advantages in that they are less liable to leak, they can be adjusted and the water cut off if necessary. They can also be used by young piglets, thus decreasing the need for a special nipple or bowl drinker for piglets.

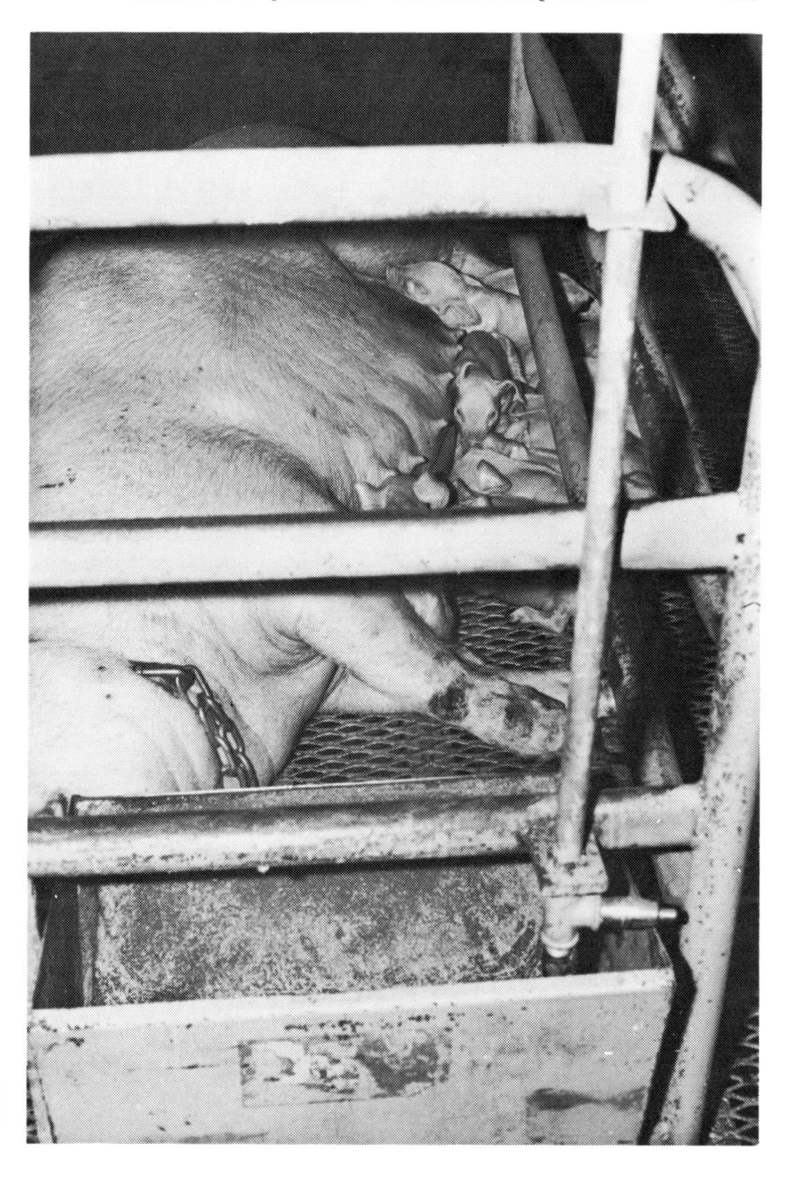

Plate No. **25**

Piglets have a strong instinct to lie as near the udder of the sow as possible in the early hours of life.

PEN SIZE

The optimum pen size is dependent on the age piglets are transferred from the pen and also, to some extent, on type of flooring, since slatted floors keep cleaner than solid floors.

The minimum desirable width for a farrowing pen is 1·8 and 1·6 metres respectively for pens with side and forward creeps. The minimum desirable lengths for pens with side and forward creeps are 2·1 and 2·7 metres respectively. If pens are too cramped, piglet losses increase and creep feeding is made more difficult.

HEATING ARRANGEMENTS

The temperature aimed for in a farrowing house is often a compromise because of the widely divergent needs of the sow and her piglets. Whilst the sow can withstand a temperature of 12°C with no adverse effects, the new-born piglet requires a temperature of 28 to 30°C for comfort and optimum performance. As the piglet gains weight, temperatures can be reduced, so that by three to five weeks of age, temperatures of 18 to 20°C are satisfactory for piglets.

Heating a farrowing house can be carried out in two ways: an attempt can be made to heat the whole farrowing house so that the requirements of the piglets can be met; more commonly, only part of the house is heated so that localised areas for the comfort of the piglets are created. In some cases a combination of the methods described are employed in the form of background heating together with a heated creep area.

Whole house or space heating

Table 7·2 gives details of the common forms of whole house heating together with their mode of operation.

TABLE 7·2

Type of heating	*Mode of operation*		
	Convection	*Conduction*	*Radiation*
Hot water pipes or radiators	√	√	√
Gas or electric heaters		√	√
Ducted warm air heaters	√		
Convector heaters (fan assisted)	√		

The amount of supplementary heat required to maintain house temperature will depend on a number of factors but principally on the temperature desired and on the ventilation rate chosen. Table 7·3 gives details of the unit costs of the various utilities used in heating. (Based on prices prevailing in UK, April 1982.)

TABLE 7·3. Unit cost of heating utilities

Utility	*Unit cost of utilities*	
	Unit cost	*Relative cost index (Oil = 100)*
Electricity	3·46p/unit (kw/h)	142
Oil 28 sec	18·46p/litre	100
Gas (Propane)	27p/kg	111

Source: Robertson, A., Scottish Farm Buildings Investigation Unit.

While these prices will change absolutely and relatively with time, the important point is that the cost per unit of heat energy from alternative utilities should be calculated as a basis for deciding on the most economical form of heating at a particular period of time.

Where electricity is used as the heat source, convector heaters or heating elements sited in the ventilation duct of a house employing a pressurised ventilation system are the most convenient form of heating. Electric radiant heaters can also be used but they are more effectively employed where localised heating is required.

Several firms market gas radiant heaters and convector heaters suitable for whole house heating. Warm air can also be supplied using an oil-fired burner.

Oil-fired heating is most commonly used in conjunction with hot water systems with unlagged pipes or with conventional domestic radiators. Table 7·4 gives details of some of the heating appliances available together with their approximate heat output.

TABLE 7·4. Heating appliances and their approximate outputs

Appliances	*Approximate heat output*
Electric heaters (fan assisted)	1–3 kW
Gas radiant heaters	0·40–5 kW
Gas convector heaters	18–100 kW
Oil-fired (unlagged pipes) (hot water)	141 Watts/m
Oil-fired (radiators) (hot water) (single panel)	600 Watts/m^2 (single panel radiator)

Control of the heat output of heating appliances should be carried out in conjunction with control of ventilation rates where these rates are subject to variation. Where the ventilation rate is held constant, control of the heat output of heating equipment can be obtained using a thermostat.

Localised heating
The most common forms of equipment supplying localised heat are gas and electric radiant heaters. Localised heat can also be supplied by means of electric warming cables and hot water pipes buried in the floor and by electric warming cables encased in rubber to form a mat which can be laid on the floor of the creep area. Localised or zone heating is necessary in order to provide the required thermal environment for piglets without the need to heat the whole house. Such areas encourage piglets to lie away from the sow when otherwise they would tend to lie in an area where the risk from crushing is high.

TYPES OF HEATER

Some relevant comments on alternative types of heating equipment are detailed below:

Infra-red lamps (electric)
Usually suspended approximately 0·45 m above the floor in the creep area. Normally 250 watts.

Infra-red heaters (electric)
Either bright or dull emitters.

Bright emitters produce emergy in the form of light as well as heat, whereas dull emitters provide heat only and require a small pilot light to attract piglets to the heated area. These normally provide 300 watts but this depends on their length. Such heaters give a better spread of heat than lamps in long narrow creeps.

Dimmer switches can be used in individual or in a series of lamps or heaters to reduce heat output as piglets grow and so economise on electricity.

These controls are becoming an increasingly worthwhile investment in view of the escalating cost of energy.

Gas heaters
These normally provide between 300 and 650 watts. Heat output can be controlled using manually operated regulators.

Underfloor heating
Heating should be provided in the piglet creep area/areas and not under the sow and can be supplied by hot water pipes or electric warming cables.

The heat supplied by hot water pipes depends on the diameter of pipe used, water temperature in the pipe and room temperature. Control of heat output can be obtained using thermostatically controlled valves on an individual pen or room principle. In pens with underfloor heating, it is necessary to provide a heat lamp or pilot light during the first 48 hours to attract piglets on to the warmed area.

Electric underfloor heating
This is similar in principle to hot water heating.

Control of heat output can be obtained using thermostatically operated simmerstats on an individual or room basis.

Heating pads
Movable heat pads are manufactured in synthetic rubber or fibre glass sheeting with a grid of warming cables embedded in the mat. The cables operate at 24 volts so a transformer is required to step down voltage from the mains.

A heat pad for a litter up to three to four weeks of age should have a minimum area of 1 m^2, with the smallest dimension not less than 0·45 m.

A covered surface temperature of 35°C should be satisfactory and, to achieve this temperature, a loading of 300–400 W/m^2 is required. Control of heat output can be achieved by means of a simmerstat type regulator.

WHOLE HOUSE v CREEP HEATING

The choice between whole house or creep heating or a combination of both is straightforward when the decision is based on capital and maintenance costs. It is cheaper to heat a relatively small area of a farrowing house than to heat the whole house. The creep area of a farrowing house can be maintained at a temperature suitable for piglets using approximately 0·30 kW/hour/sow and litter, whereas whole house heating may require between two and three times this amount of energy. If creep heating is used, it must be remembered that heat lamps or any other form of creep heating can set up air currents which can chill piglets, especially where the rest of the house is somewhat cold. Providing a covered creep will reduce such draughts and, at the same time, help to conserve heat and create a cosy micro-environment for the piglets.

SUGGESTED FARROWING PEN LAYOUT

As stated at the outset, it is impossible in planning a farrowing pen to cater fully for all desirable objectives since some are antagonistic to others. In particular, it is difficult to marry together effectively provision for the needs of the baby piglet and maximum labour economy.

The point has already been made that all farrowing pen arrangements are a compromise between often competing and contrasting objectives. For instance, there is particular difficulty in resolving where the balance of the advantage lies between pens with forward creeps (and a front service/inspection pass) and pens with a side creep (and a rear service/inspection pass). It is recognised that pens with both a front and rear pass possess greater advantages in that it is possible to inspect the hind end of the sow and feed from the front pass without having to step into the pen, but the cost of an additional pass is difficult to justify in most circumstances.

If there is to be only a single pass, we consider that the balance of advantage comes down in favour of a rear pass and, if approximately the same amount of space is available, that a pen with a side creep is preferable to one with a forward creep. Our assessment has been based mainly on the need for regular inspection of the sow from behind around farrowing, so as to detect problems promptly and to facilitate any treatment of the sow and assistance measures for piglets without causing undue disturbance to the sow; the preference for the side over the forward creep is designed to cater for the piglets' instinctive preference to lie as close to the udder as possible in the first few hours of life.

That is why, in the recommendation made in Figure 7·2, firm emphasis has been placed on providing an adequate opportunity to supervise farrowing without causing undue disturbance and on catering for the welfare of the newly-born piglet. Although this may involve a slightly higher labour requirement than some alternative designs, it is our assessment that the benefits of reducing early piglet losses are likely to more than compensate for marginally higher labour costs now and in the foreseeable future.

The particular features of the farrowing arrangement suggested in Figure 7·2 are as follows:

Floor

(a) There is a substantial area of concrete at the front of the pen to provide a comfortable lying area for piglets and to facilitate creep feeding. The concrete area should be sloped towards the slatted area so that it is kept as dry as possible. The remainder of the pen is slatted using either woven wire or welded metal. Such a floor provides a comfortable bed for the sow and is likely to encourage adequate exposure of teats when nursing.

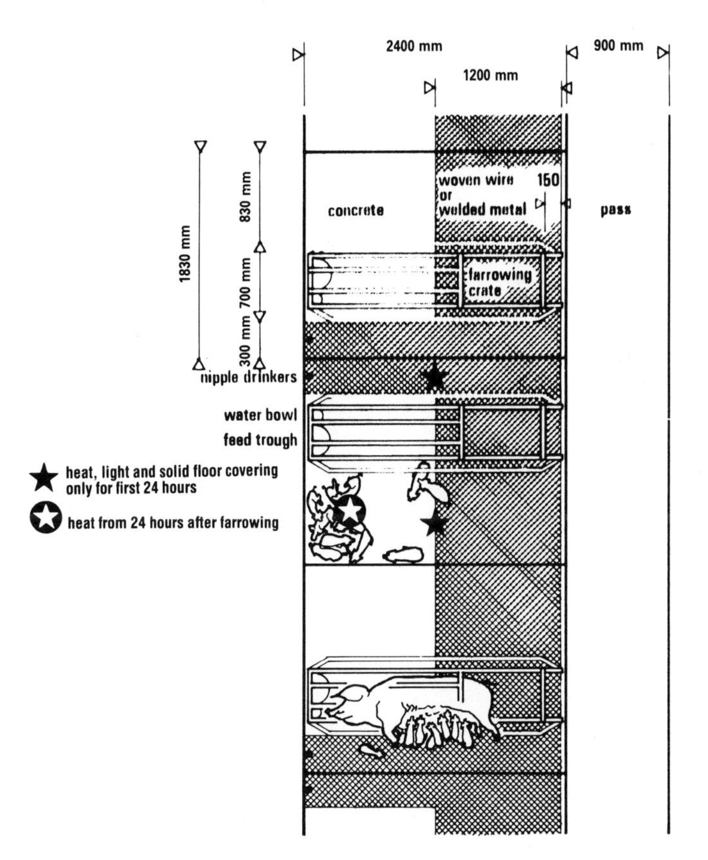

Figure 7·2. Recommended pen layout for farrowing.

(b) A 0·6 to 1·2 metre strip of welded metal or woven wire at the rear of the pen for cleanliness. These types of slatted floor have been shown to minimise risk of injury to the udder and piglets, provide for good piglet mobility, good foothold by the sow and to keep clean. If necessary, the amount of the floor covered with these types of flooring can be increased although it is desirable to have a reasonable amount of solid flooring as this allows bedding to be used to help maximise comfort for the piglets when they are most vulnerable in the first few days of life.

(c) A 0·3 metre wide strip of slatted floor running along the pen in the narrow creep to cater for the piglets' dunging/urination in that area and to provide drainage from the piglet nipple drinker. The solid area of the floor should be sloped gradually towards the slatted areas. This narrow strip of slat down the side of the pen may be disposed of to facilitate construction and the piglet nipple mounted on the farrowing crate over the slatted area at the rear of the pen. When a water bowl is used for the sow to which piglets have ready access, the piglet nipple drinkers are dispensable. However, one advantage of installing piglet nipple drinkers is that it provides the necessary training for piglets when this type of drinker is to be used after weaning.

Crate

The farrowing crate arrangement recommended for the suggested farrowing facility is that suggested in Figures 7·1(b) and 1(c) (page 147) with the addition of farrowing cradles as illustrated in Plate 23.

This arrangement assists the sow in controlling her descent from the standing position and obliges her to lie down on her belly before rolling over to either side. Thus, she is prevented from flopping directly on to her side.

Crate offset in pen

The farrowing crate is offset in the pen with the narrow creep being 0·3 metre wide and the other creep 0·8 metre wide. This wide creep provides adequate room for lying on the solid-floored part of the pen and also for creep feeding. By using a solid board along the side of the crate, catching of the pigs is also simplified.

Drinking and feeding arrangements

The sow feed trough is mounted on the floor and a self-fill water

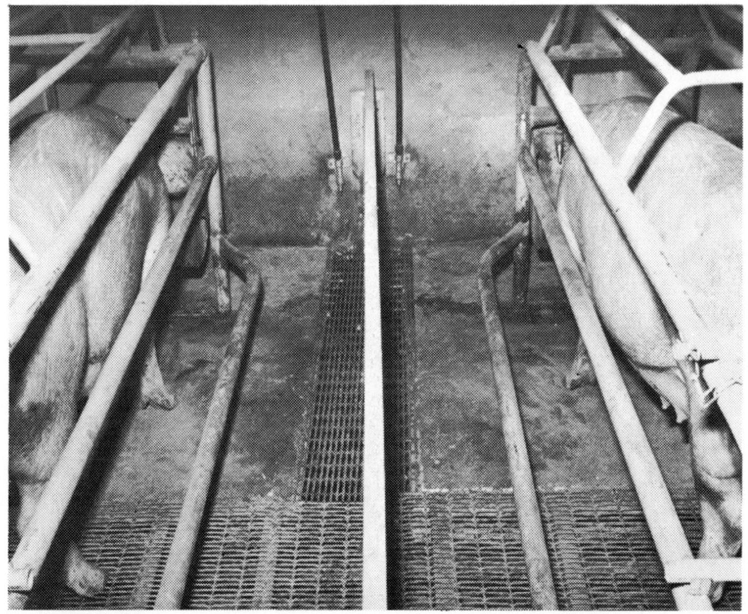

Plate No. **26**

A narrow strip of slatted area down the narrow creep helps to keep this area clean and dry and provides drainage for nipple drinkers, if provided.

bowl is adjacent. This bowl can be shared by sow and piglet which eliminates the need for a piglet nipple drinker unless such a drinker can be justified for reasons of training piglets to use nipple drinkers after weaning.

Heating

Heating can be organised using any of the alternatives discussed, depending upon initial capital cost, running costs, convenience, reliability, efficiency and such factors. It is desirable that heat and light be available in both side creeps in the first 24 hours so that piglets will be attracted away from the udder when suckling, whichever side the sow lies on. When heat is provided only on one side, there is a tendency for the sow to lie with her back to the heat so that newly-born piglets will be suckling on the cold side. Hence the importance of a heat source on this narrow side in the first 24 hours; a paper sack or other solid covering can be placed above the mesh to increase piglet comfort in this period.

A further useful modification is to have one of the heat lamps swivelled round to the area adjacent to the vulva for farrowing to increase comfort at the place of birth. This is a particularly useful practice which will be of special assistance to smaller, weaker piglets at birth if the farrowing house is colder than it should be for any reason. After farrowing has been completed the lamp should be swivelled round to the side creep area. It is desirable that the heating device is controllable so that heat output, and therefore costs, can be decreased as the piglets grow.

Rear service/inspection passage

A rear service/inspection passage is provided for the entry of sows

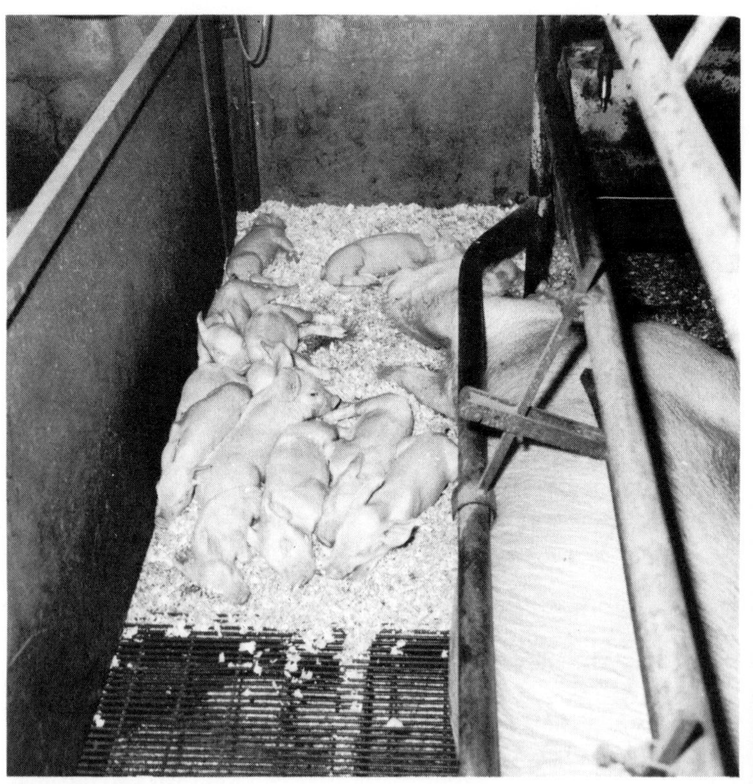

Plate No. **27**

An adequate environment in the farrowing house and in the creep area, in particular, is of vital importance in getting the maximum number of piglets established. The piglets' behaviour is the best indicator of their comfort. Those above are nicely spread out in the creep and appear to be in perfect equilibrium with their environment.

and to facilitate checks during farrowing and provision of any assistance in getting smaller piglets established on the udder soon after birth.

Creeps adjacent to each other
The wider creeps in neighbouring pens should be adjacent to each other so that, after the first few days, heat is concentrated in a fairly restricted area. This arrangement also means that the narrow slatted areas down the side of neighbouring pens are adjacent, allowing one 0·6 m slat to serve two adjacent pens, thus decreasing building costs.

Adjustable crate fronts
A proportion of crates in the farrowing house should have adjustable fronts to ensure that as much as possible of the effluent of the smaller gilt is dropped on the slatted area, in pens where there is a narrow slatted area (0·6 metre or less) at the rear of the pen.

Controlled lighting
Controlled lighting should be provided so that the house can be kept dimmed most of the time, thus reducing restlessness of sows. Where heating lamps are used, extra lights need only be used at such times as feeding.

GENERAL COMMENTS

If the pen width is some 1·8 metres, this pen arrangement should cater for litters up to four to five weeks of age.

In summary, the pen arrangements aim at:

(a) providing maximum comfort for the udder of the sow and the piglets;
(b) encouraging the sow to expose effectively all teats on the lower row of the udder at suckling;
(c) providing a floor surface within the danger zone for crushing which caters for good mobility in piglets which will help them to escape from danger, and good foothold by the sow;
(d) encouraging piglets to lie from soon after birth in a heated zone very adjacent to the udder but still outside the danger zone for crushing;
(e) providing strategic slatted areas to cater for the dunging/urinating habits of the sow and piglets and so leading to cleanliness and good hygiene within the pen. In addition,

the risk of the lying areas being wetted by accidental spillage from nipple drinkers is minimised;

(f) providing adequate space for creep feeding.

FOLLOW-ON PENS

For those practising five- to eight-week weaning who wish to have a specialist farrowing facility such as that suggested in Figure 7·2 and who, for various reasons, wish to transfer to follow-on accommodation after 10 to 20 days, there is an infinite variety of such accommodation available. If possible, this transfer should be delayed until at least two weeks after farrowing and care should be exercised in ensuring that follow-on accommodation is adequate so as to avoid losing a great deal of the advantage likely to be gained from having a purpose-built maternity facility.

CONCLUSION

Many important objectives have to be kept in mind when considering the design of farrowing facilities. Some of these objectives are antagonistic to each other. It is strongly recommended that the primary emphasis be placed on the adequate welfare of the sow and new born piglets and that after their requirements are met, as much consideration as possible is given to other important aspects such as capital cost and labour economy. The primary emphasis on the farrowing sow and new born piglets is not only important from the welfare aspect but also because of the importance of an adequate farrowing facility in minimising piglet losses and in allowing the maximum number of new-born piglets to get established. One farrowing pen caters for as many as 8 to 12 farrowings per year, depending on the age at weaning and timing of transfer from such quarters, and one does not have to save many extra piglets in each farrowing pen in a year to cover slightly higher capital expenditure and labour costs. The 'extra' or 'opportunity' pigs saved by having a purpose-built maternity/nursery facility leave a large profit margin over the extra feed costs incurred in rearing them.

It might help to alter the philosophy of pig farmers if the term 'farrowing quarters' was changed to 'maternity/nursery quarters'. This might encourage them to put the emphasis in the right place for the benefit of both the piglets and their finances.

It is vital for both piglets and pig farmer that the maternity/nursery facility is designed to ensure optimum comfort, safety and nutrition for the piglets. Such conditions will help to get the maximum number of piglets established on the road to profit and the minimum number of small, pitiful bodies consigned to the dung pit.

Chapter 8

THE UDDER: SUCKLING AND NURSING

THE UDDER of the sow, in providing colostrum and milk, constitutes the sole source of nutrition and protection against prevailing infection for the piglet in the first two weeks of life. The antibodies which the sow has produced to protect both herself and her piglets against prevailing infection cannot pass through the placenta to the unborn piglet. Colostrum, or first milk, is rich in such antibodies and provides piglets with excellent protection against infection prevailing in the herd. Colostrum changes gradually to milk over the first few days and these products constitute the sole food of piglets in the early part of life.

Thus, to be well protected against disease and to be well fed, the sow must have adequate colostrum and, later, milk available and every piglet must have the ability and opportunity to obtain these products from the sow.

Many factors can influence the availability of colostrum and milk and many others can deny piglets the opportunity to obtain adequate colostrum and milk from the sow, even when it is available. These factors must be recognised so that they can be controlled and thus piglets provided with their full requirements.

The basic structure and faults of the udder will be outlined before remedial measures are discussed.

STRUCTURE OF THE UDDER

The udder of the domestic sow usually consists of between twelve and sixteen mammary glands or mammae. A typical gland is depicted in Plate 28. The actual glandular or milk-producing tissue of any one mamma is subdivided into two separate parts, each part having its own independent canal carrying the secreted milk to the teat. And so, when a teat is squeezed, milk may appear from two small openings. These openings are the ends of the tubes collecting the milk from the separate parts of a particular mamma. It is possible for one part of a mamma to become non-functional through injury or disease and another part to be perfectly normal. Of course, when a part of a mamma becomes non-functional, overall milk production of that mamma will decline.

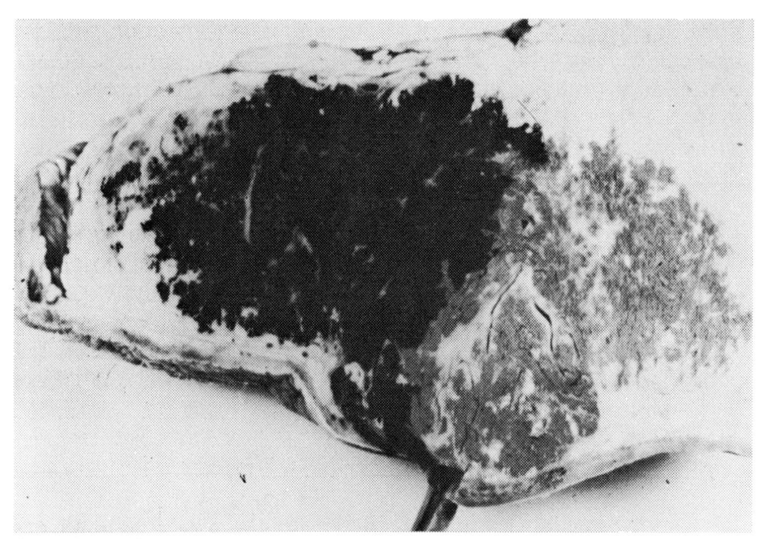

Plate No. **28**

Cross-section of a mammary gland. A special staining process has been used to demonstrate that a so-called 'mammary gland' consists of two independent glandular parts.

By courtesy of Jones, J. E. T., London Veterinary School

Thus the separate mammary glands may be:

1. Fully functional (both glandular parts operating).
2. Partially functional (only one glandular part operating).
3. Non-functional (both glandular parts damaged or diseased).

TEAT AND UDDER ABNORMALITIES

Particular faults in the udder and functional teats are covered later in this chapter. Before considering faults in functional teats, the importance of avoiding non-functional teats in selection of replacement stock must first be noted. Among the abnormalities found are the following:

- Blind teats.
- Inverted teats.
- Extremely short, blocky teats.

Blind and inverted teats are non-functional in that milk cannot be withdrawn from them by piglets. Extremely short, blocky teats are, in theory, functional, but piglets may find it impossible to grip these teats adequately in order to obtain an adequate suckle. Therefore, in practice, such teats may, in fact, be non-functional. If potential breeding gilts have been reared on rough concrete, some teats may have been partially sloughed off through excessive abrasion and these too may be largely non-functional because they cannot be adequately gripped by piglets during suckling (see Chapter 3, Teat Necrosis).

It is important in relation to maximising opportunities for nutrition, and therefore survival, that gilts should have fourteen to sixteen functional teats. Existence of teat abnormalities such as those noted above effectively reduces the theoretical rearing capacity of gilts and sows, and for this reason potential breeding animals should be checked carefully for the presence of these abnormalities.

DEVELOPMENT OF 'TEAT ORDER'

Very soon after birth piglets compete for the available teats and the associated mammae, and, judging from the intense competition, some teats are obviously more desirable than others. Once the 'teat order' has been settled, piglets tend to retain their selected teats up to weaning.

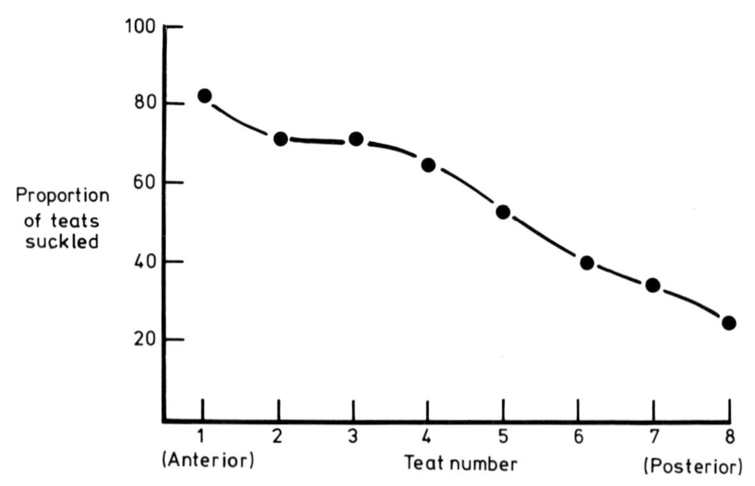

Figure 8·1. Proportion of teats suckled in various udder positions (based on 104 sows).

It is clear from Figure 8·1 that the front or anterior teats are more popular and a higher proportion of these are suckled.

FAULTS IN THE 'REARING MACHINE'

Among the very important characteristics of an efficient machine for rearing young animals are either individual feeding—with teats or troughs placed so that each animal is ensured its share of feed—or else incorporation of exactly the same characteristics in each feeding position so that animals will move happily to different positions rather than compete with each other for favoured spots. As Figure 8·1 indicates, the sow's udder does not possess such desirable characteristics as the front teats are selected in preference to the rear ones. This has a serious effect on piglet mortality because of the fighting involved when the 'teat order' is being developed, and because sows let down their milk once per hour and milk flow lasts for about 20 seconds only.

That the front teats are more desirable to piglets means that soon after birth piglets try to gain a front teat. The net result is often a scramble for front teats as piglets are 'grunted up' by the sow and this may result in the 20-second milk flow passing before quarrels are settled and teats are adequately gripped. Since a newborn piglet has a low energy reserve, the missing of a feed can be crucial, while missing several feeds in succession is often disastrous. With each successive feed missed, the chances of ensuring an adequate suckle next time, and of survival, are progressively reduced. The situation is made even worse if the environment is sub-optimal.

In theory, it would be much better if all teat positions on the udder were equally desirable. The mad scramble for the front teats would be avoided, piglets would spread themselves evenly along the udder and they would be more likely to take full advantage of the very brief milk flow as soon as it began.

If the differential popularity of feeding positions applied to an artificial rearing device for group feeding, manufacturers would very quickly undertake exhaustive studies so as first to isolate and then eliminate, the causes. It does seem strange that although the differential popularity of the sow's teat positions has been known for a considerable time, it has been accepted and taken for granted, and no scientific work has been applied to put the problem right. It may be that the true significance of this deficiency in contributing to piglet mortality by increasing competition for suckling positions within the litter has not been fully appreciated.

Before any attempt can be made to achieve more equal popularity of different teat positions, we must try to isolate reasons for the differences which exist.

WHY ARE FRONT TEATS MORE POPULAR?

Front teats may be preferred because of greater security—a kick by the front leg may carry less threat and pain than one from the hind leg! Moreover, piglets massaging the front teats may be more successful at bringing on milk 'let down' by the sow than those suckling the back teats. These factors may make front teats more popular.

However, some additional characteristics which differ with teat position and which may be modifiable by appropriate selection have been isolated in our work.

It can be seen from Table 8·1 that there is a general tendency for spacing between adjacent teats to decrease progressively from teats 1 to 6. There tends to be a wide spacing between the penultimate and last teat. An almost identical pattern was found in ten other herds examined. Reduction in spacing between teats from front to back (the last teat excepted) may reflect differences in the amount of glandular tissue associated with each teat position. If this is so, it may explain some findings which indicate that front teats are more productive than back ones.

Table 8·1. Mean spacing between adjacent teats

	Teat number (front to rear)					
	1–2	2–3	3–4	4–5	5–6	6–7
Mean spacing (cm)	10·5	9·9	9·6	9·4	8·0	12.4

(Based on 57 sows)

Another characteristic which shows consistent changes from front to back of the udder is teat length (see figure 8·2), which is greater in the front positions. It would seem logical that a piglet would prefer a long slender teat to a short stubby one to enable it to obtain a better grip during suckling. If so, then an index of teat length to teat diameter may be a useful indication of popularity. Our work (see Figure 8·2) showed that there is a clear tendency for front teats to be longer and more slender than rear teats.

Another reason for the greater popularity of front teats is the fact that if the orientation of teats on the lower row as presented to piglets at suckling is examined, it will be found that the front teats are more available. This is shown in figure 8·3.

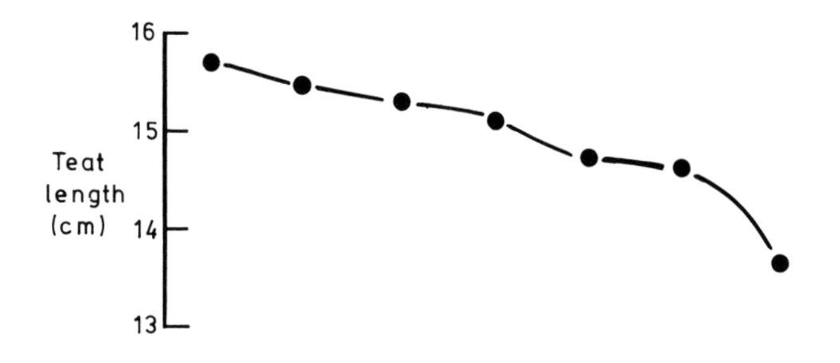

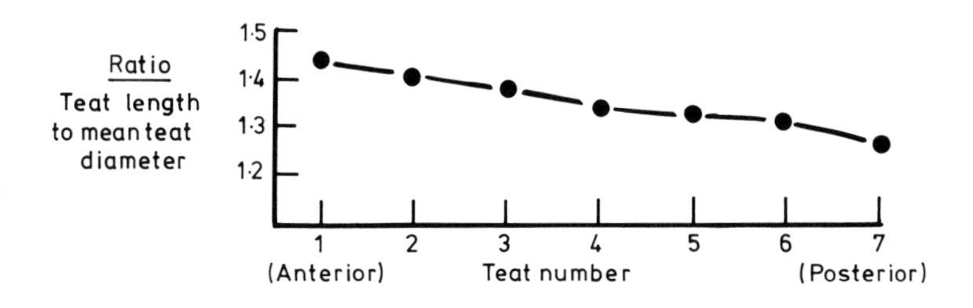

Figure 8·2. Teat dimensions according to position on udder (based on 94 first lactation sows).

Thus, front teats have more clearance above floor level than rear teats and, the greater the clearance, the more effectively piglets can grip teats. It will be noted that clearance declines up to teat six. The hind-most teat usually has greater clearance above floor level than those teats immediately anterior to it. It will also be noted that average clearance of teats above ground floor level deteriorates as sows age. The udder becomes more pendulous as sows age and this decreases their ability to expose the bottom teats effectively at suckling. It must also be noted that Figure 8·3 is based on average figures.

The front teats of some sows, of course, will have a greater clearance above floor level than the average and the clearance will be less in others. In the same way, the rear teats of some sows will have a greater clearance, while those of other sows with more pendulous udders will have no clearance above floor level at all, i.e. they will not be exposed effectively to piglets at suckling at all. This problem of failure to expose all rear teats on the lower row is discussed in the next section.

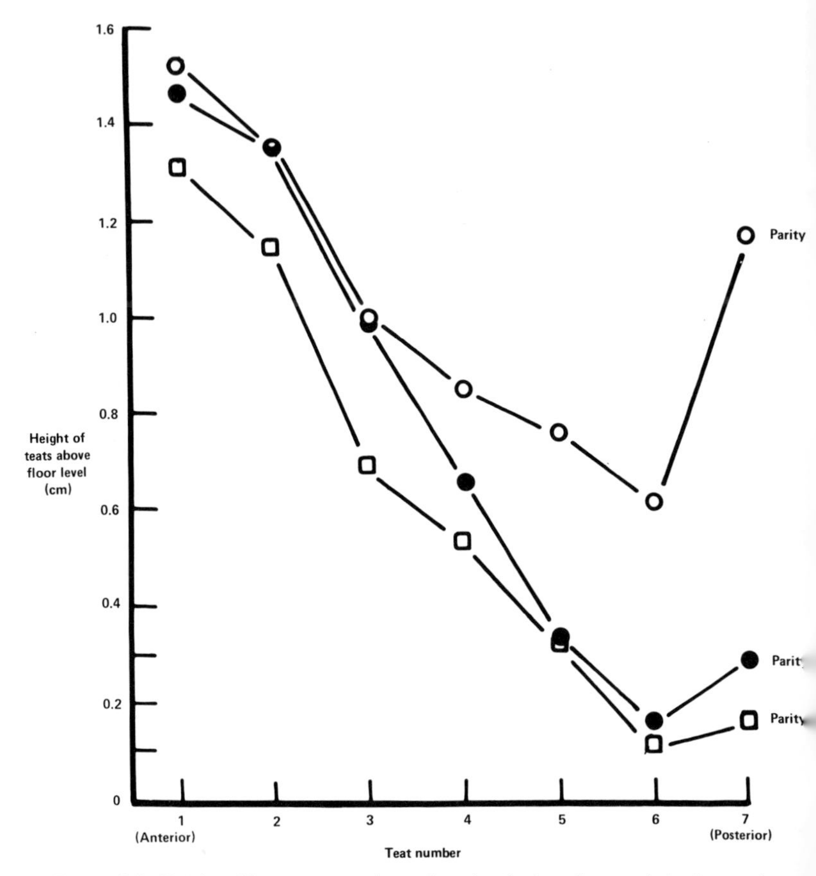

Figure 8·3. Height of bottom teats above floor level when the sow is in the nursing position (averages based on 66 sows).

It is shown, therefore, that udder characteristics such as teat length, diameter and shape, spacing between teats and degree of exposure of teats vary along the length of the udder and it is likely that these physical factors play an important role in the greater popularity of the front teats. In selection of gilts, some attempt should be made to decrease the differences existing in such factors as teat shape and teat spacing along the length of the udder in an attempt to increase the acceptability of the rear teats.

Differential popularity of teat positions must be reduced if competition between piglets for teats, and consequently piglet mortality, are to be minimised.

Some pig-keepers who are aware of the tendency for close

spacing of the rear teats make a very conscious effort to avoid this since they appreciate its effect, not only on piglet growth, but on mortality. Wider teat spacing keeps a little pig's worst enemy (another little pig) that bit further away from it when it is trying to get its full ration of life-giving colostrum or milk.

'TEAT ORDER' AND PIGLET MORTALITY

That one piglet selects a given teat and retains it throughout suckling at each of the 20-second milk 'let down' periods means that there is no opportunity for two piglets to share one teat. If such confrontation does arise, it is very much a question of the survival of the fittest.

However, if sows have been carefully selected, they will have fourteen to sixteen functional teats and these will be sufficient to cope with most litters. For litters with piglets surplus to the number of functional exposed teats, alternative provision must be made for the suplus piglets.

While many sows are effective in exposing all the teats on the lower row at suckling, others are ineffective in exposing the more posterior teats on the lower row (see Plates 29A and 29B).

The sow in Plate 29A exposes all teats on the lower row effectively to her piglets at suckling. However, the sow in Plate 29B, although she has seven functional teats on the lower row, succeeds in exposing only three of these at suckling. The remaining four teats are hidden under the udder. Although this latter sow has fourteen functional teats, only ten of these are being made available to piglets and thus the sow can cope effectively with a maximum of only ten piglets. This problem of failure to expose the rearmost teats on the lower row at suckling is more common in older sows with more pendulous udders. Some 40 per cent of such sows may fail to expose the last three to four teats on the lower row at suckling. Surprisingly, as many as 20 per cent of first and second litter sows can show a similar fault. Possibly because of some post-farrowing discomfort, sows are often seen to experience particular difficulty in adequately exposing all bottom teats soon after farrowing. This time is, of course, crucial. Piglets surplus to number of exposed teats at this stage will soon fade away through starvation.

This problem limits the *rearing capacity* of sows. Rearing capacity is not determined by the number of functional teats on sows but by the number they expose effectively to piglets at suckling in the first crucial one to two days of life. Piglets in excess of the number of functional exposed teats at this time are surplus

Plate No. **29A**

Well-exposed udder at suckling with all teats readily available to piglets.

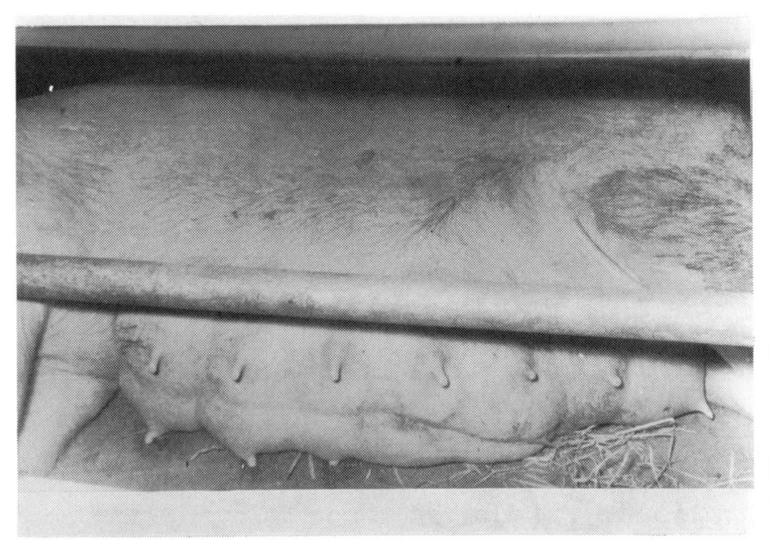

Plate No. **29B**

Poorly exposed udder at suckling with the four most posterior teats on the lower row unavailable to piglets. This problem reduces the 'rearing capacity' of the sow.

to rearing capacity (supernumerary piglets), and alternative provision must be made for these extra piglets by, for example, fostering to sows farrowing at the same time which have spare rearing capacity available.

It is difficult to suggest a suitable solution to the problem of failure to expose the rearmost teats on the lower row at suckling. Since this problem is greater in older sows, timely culling of sows which show this tendency will obviously help. Attempts to provide a raised platform in the middle of the farrowing pen so that the lower row of teats just overhangs the edge of the platform at suckling have not been very successful, mainly because one cannot control the exact position in which the sow will lie. In addition, with the sow on a raised platform, some teats on the upper row may be relatively inaccessible to piglets at suckling. Selection of sows with the two rows of teats closer together is a suggestion which merits further consideration for decreasing the problem. It is likely also that the more comfortable the floor surface for the sow, the more willing she will be to rotate her body fully when attempting to present the maximum number of her teats on the lower row to

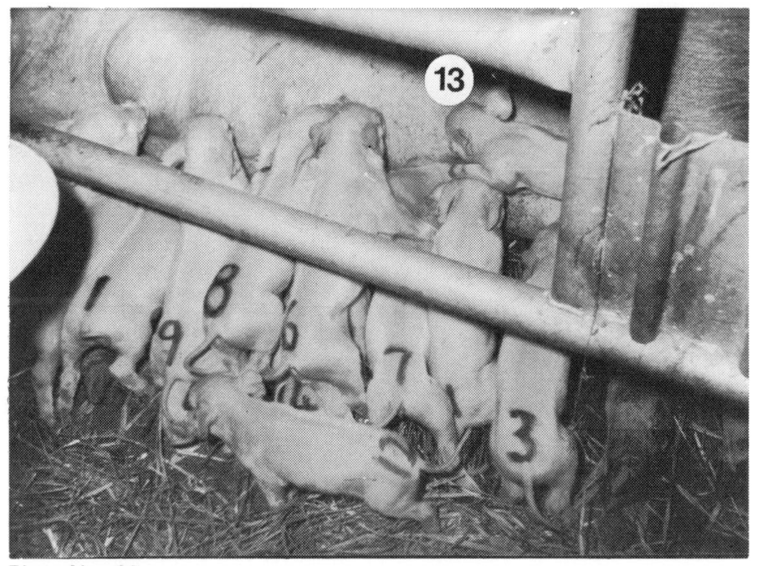

Plate No. **30**

Piglets 12 and 13 are supernumerary in this litter i.e. they are surplus to the number of exposed functional teats on the sow. Finding a suitable foster sow with spare 'rearing capacity' promptly is the only practical solution to this problem at the moment.

her piglets at suckling.

Enough has been written to show that the chances of a normal piglet, which is vigorous at birth, surviving is not simply dependent on there being enough functional teats on the sow to go round. Such factors as unavailability of the back teats on the lower row and the tendency for piglets to cluster at the front of the udder—sometimes at the expense of leaving rear teats vacant—must be taken into account. If piglets are in a similar plight to No 12 and No 13 (in Plate 30) then obviously a foster dam must be found. In the case of litters which tend to cluster at the front of the udder, with more posterior teats being neglected, then sometimes piglets which are apparently unsuccessful in obtaining a teat can, with a little encouragement from the stockman, be persuaded to accept a more posterior teat. If this is unsuccessful, fostering the pigs is the only solution.

The sooner piglets surplus to rearing capacity are transferred to a foster dam, the better for maximising their chances of survival. Such prompt action also improves the survival chances of their litter-mates, for as long as surplus piglets are around and active, they interfere at each feeding time with the remainder of the litter and they often run around in a frustrated manner, issuing high-pitched squeals which can upset the composure of their dam.

The problem, then, is simply to isolate those piglets which are seemingly unable—for any of the above reasons—to obtain a teat and get an adequate, regular suckle. These *supernumerary or underprivileged* pigs can be readily spotted at any of the frequent feeding periods and it is vital that the stockman intervenes at a very early stage and makes appropriate alternative provision for rearing before piglets become excessively weakened. Timely action of this kind can successfully convert a would-be skeleton into a money-spinning commodity.

Need to transfer piglets which are surplus to their dam's rearing capacity creates a strong requirement for batch-farrowing to increase the opportunities for fostering. The need for a simple and successful 'artificial sow' to cope with such surplus piglets when there is no opportunity for fostering is also brought into focus. These practices are discussed in greater detail in Chapter 9.

TEAT ORDER AND GROWTH RATE

It would simplify management a great deal if pigs in a litter were like 'peas in a pod' at weaning—identical in size and condition. There would be no need for mixing litters to get even groups for the finishing pens with consequent risk of growth checks.

Plate No. **31**

Piglets suckling the most posterior teats often grow less than those suckling front teats.

We know that the very even litter is the exception rather than the rule. The sow has let us down again! Why? There are various reasons why suckling piglets do not grow equally well. Among these are:

- Differences in birthweight within the litter.
- Differences in milk production between the feeding places of the 'rearing machine', the udder.

When litters are suckling, it can often be seen that piglets on the teats towards the rear tend to be smaller (see plate 31).

This trend is also evident in Table 8·2 which relates piglet weight to teat number suckled.

TABLE 8·2. Weight in relation to teat number suckled (kg)

	1	2	3	4	5	6	7
Weight at birth	1·41	1·29	1·31	1·32	1·34	1·31	1·25
6 weeks	10·7	9·7	9·2	8·7	8·8	8·2	8·5

(Based on 104 litters)

It can be seen that there is a tendency for the larger piglets at birth to select the front teats and for the smallest piglets to be relegated to the rearmost teat.

However, there is a wide range in weight at six weeks between teat positions which is not explained by birthweight; for example piglets on teats two to six are very similar in birthweight but show a considerable range in weight at six weeks of age.

This indicates that the glands towards the rear of the udder produce less milk and this may be associated with the gradual reduction in spacing between teats one to six (see Table 8·1) and an associated reduction in amount of glandular tissue from the front to the rear of the udder.

It is likely that the variation in weaning weight associated with teat position suckled could be reduced by paying more attention to the underline in gilt selection, in particular to even spacing between teats and uniform teat development along the length of the udder.

But, if we were being much more selective on underline, we would be rejecting some gilts for breeding which were very good on other traits, such as weight for age and carcase quality. Some may prefer, therefore to put up with the defects in the sow which lead to variation in the growth of piglets within the litter and make alternative provision for any piglet or piglets which are falling considerably behind the rest.

NUTRITIONAL 'RUNTS'

One can have variation in growth, and therefore size, of piglets within a litter but with all piglets thriving reasonably well. On the other hand, one can have a litter in which most piglets are thriving well but in which one or two are obviously suffering from malnutrition and are in danger of becoming nutritional 'runts'.

Such ill-doing piglets are most noticeable in large litters. All piglets may get off to a good start, but around ten days or so, increasing variability in the litter is often apparent. Perhaps ten piglets in the litter may be doing perfectly well but the remaining one or two appear to be losing condition relative to the rest, either because all cannot physically get into the udder at one time for suckling and the smaller, weaker ones are pushed out or else the teats claimed by these piglets may have dried up or be very unproductive. It may well be that their teats are not fully functional, i.e. only one of the separate glands associated with that teat (see Plate 28) may be functioning. This may be able to provide sufficient milk when piglets are small and nutritional demand low,

Plate No. **32**

As a litter grows, some piglets experience increasing physical difficulty in gaining access to the udder at suckling.

but is incapable of providing for the increased demands as the piglet grows older.

However, whatever the basic cause, this problem is a fairly common one especially in herds weaning at five to eight weeks of age which tend to have higher numbers born per litter. These undernourished piglets may die before weaning or else end up as 'runts' at weaning.

The problem of these potential nutritional 'runts' may be tackled in a variety of ways:

- They may be early-weaned and transferred to an artificial rearing system based on dry feeding as soon as they are seen to be falling behind their litter-mates. This practice is termed 'complementary rearing' and is discussed in more detail in the following chapter.
- An attempt can be made to foster these piglets on to a newly farrowed sow with spare rearing capacity. This is usually successful but on occasions can jeopardise the performance of some newly born pigs in the host litter.

- Eight to ten of these undernourished piglets may be gathered together and fostered on to a newly weaned sow which is still milking reasonably well. The advantage of this system over the early-weaning and 'complementary rearing' approach is that no further capital investment is required, and the sow rather than the stockman is saddled with the work of rearing. On the other hand, the disadvantages of being dependent on such a sow are that one may not be available just when required and good milking sows on five- to eight-week weaning systems may already have been brought well down in condition. In addition, the sow may refuse to accept the piglets without being tranquillised. However, some producers use this system successfully and, as a foster sow, use either one about to be culled or, alternatively, stimulate the foster sow to take the boar while still suckling so as to minimise loss in litters per year.

Whatever system is adopted for dealing with these piglets which show signs of malnutrition from seven to ten days of age, it is vital that these piglets be detected promptly and alternative nutrition arranged without delay. There is no point in waiting until these piglets have lost a lot of condition before taking the necessary action.

MILKING ABILITY

So far, only the structure of udder and teats have been discussed. Structural faults were isolated which tended to increase piglet losses and lead to variation in piglet growth within litters. Suggestions were made regarding the improvement in structure of the udder and teats with a view to improving piglet survival and achieving more uniform growth of piglets within litters.

However, ability to produce an adequate supply of milk is as important as good structure of the udder and teats. Milk production from well-structured mammary glands is dependent on:

- The genetic potential.
- The provision of adequate nutrients as a basis for production of the constituents of milk.
- The health of the milk-producing glands and of the sow in general.

Provision of adequate nutrition (energy, protein and amino acids, vitamins, minerals and water) to meet demands for milk production is covered in detail in Chapter 11.

Regarding genetic potential, it is recognised that differences in milk production capacity exist between breeds and crosses and between individuals of the same genotype. The most useful practical index of milk yield of the sow is the weight of the litter at three weeks of age, since up to this stage, piglets show little interest in solid food and so their growth is almost entirely dependent on the milk yield of the dam.

A sow which is well fed in relation to providing for milk production may fail to milk because of agalactia, this being one of the consequences of post-farrowing fever.

POST-FARROWING FEVER–CAUSES AND REMEDIES

This condition is usually referred to as the MMA syndrome. These letters stand for the following, M = mastitis (inflammation of the udder), M = Metritis (inflammation of the womb), A = Agalactia (no milk production). The MMA syndrome is a complex condition involving metabolic, bacterial and hormonal factors with stress playing a part. The most obvious and serious symptom is either partial lactational failure (hypoagalactia) or complete lactational failure (agalactia).

The clinical signs include increased respiratory rate, increased heart rate, depression, inappetance, increased temperature, constipation, reluctance to rise, failure to expose teats and nurse, mastitis of one or more glands, blotchiness of the skin and vaginal discharge; all these signs may be present but usually various combinations of some are present. A vaginal discharge does not necessarily indicate the presence of metritis and can be present in apparently normal sows after farrowing.

Workers in various countries have noted that: (1) The condition is more common in old fat sows fed on unbalanced diets and which experience difficulty in rising. (2) There may be inadequate development of the mammary gland, mainly in small fat gilts reared on too liberal feeding. (3) The condition is absent from properly run 'Roadnight' or outdoor system herds. (4) The condition is more common in herds where animals are confined for most of their lives. (5) there is no immunity to subsequent attacks. (6) The disease may suddenly appear in a unit and affect every sow or gilt which farrows or it may take a chronic form affecting only an odd sow or gilt. (7) The condition may suddenly disappear from a unit without preventive measures having been taken.

Predisposing factors
Lack of exercise, overfat condition, finely ground diets and stress are factors which have been incriminated by many workers. Sows affected with the MMA syndrome do not die and will recover in the absence of treatment in three to four days. However, the condition must be promptly treated as the piglets would soon die of starvation (hypoglycaemia).

Treatment and control
The most useful step the stockman can take to help lessen the incidence and effects of the MMA syndrome is to keep a regular check on appetite, rectal temperature and piglet behaviour and well-being after farrowing. Restless, squealing piglets often indicate dissatisfaction due to the shortage of milk and this sign along with the general appearance of the piglets, occasionally elevated rectal temperature (above 40·5°C) and loss of appetite should be taken as individual or collective signs that all is not well and prompt veterinary attention should be sought.

Treatment of the sow
Affected sows may be treated with a variety of antibiotics. The veterinary surgeon attending the unit will know which antibiotics are likely to have more effect. Injection of a cortisone preparation along with the antibiotic seems to have great value. Fractious or excitable sows will benefit from sedation. Oxytocin or pituitary extract should also be administered initially. This drug causes contraction of the muscles surrounding the milk ducts, forcing the milk towards the teat. Its effect is only transient and it will only succeed if there is milk in the gland in the first place. Injections of Oxytocin should be repeated at least every two hours until such time as the sow shows evidence of recovery.

Immediately after injection, attempts should be made to encourage the sow to lie over and suckle the piglets. In some cases, affected sows deliberately lie on the udder, which rapidly becomes hard and painful to touch. Some sows may be either sedated or forcibly held with ropes. The latter device is useful until veterinary help arrives. Initially, the sow will protest violently but once the initial congestion of the udder has been relieved she will nurse the piglets normally.

TREATMENT OF THE PIGLETS

Starved piglets will rapidly become weak and comatose unless they are provided with a source of readily available energy. This may be provided in the form of artificial milk, cow colostrum or dextrose (glucose) solution. Occasionally a litter can be transferred to a newly weaned sow until the dam shows signs of recovery. Fortunately, most piglets will have received some colostrum from the dam as the disorder does not usually appear until ten to fourteen hours post farrowing. Piglets are often affected with colibacillosis when the MMA syndrome is present in the dam.

PREVENTION

If the predisposing factor or cause can be determined with certainty, specific steps can be taken to avoid the problem. However, in the majority of outbreaks the cause or predisposing factor will not have been pinpointed. Apart from specific advice from a veterinary surgeon, a series of (trial and error) changes can be made to the management or nutrition. Providing a more coarse open ration, adding molasses, providing silage or vegetables, and two ounces of Epsom salts the day before parturition are recommendations which have led to success according to some workers.

In some circumstances, reducing feed level just before and after farrowing is reputed to help reduce incidence of this condition. However, in other situations, such a practice does not appear to help and excessively low feed levels around farrowing can increase the restlessness of those sows with healthy appetites and result in higher losses from crushing in piglets.

Vaccination of sows with an autogenous vaccine made from the *E. Coli* which can be readily cultured from the vaginal discharge has successfully terminated many outbreaks. Prompt treatment of both sows and piglets is essential if losses are to be avoided from this condition.

MILK COMPOSITION

The first milk or colostrum has a very high solids content, the protein fraction being particularly high. The globulin fraction of the protein contains the antibodies which protect the piglet against infection prevailing in the herd. Composition changes very quickly after farrowing to that typical of sow milk as shown in Table 8·3.

TABLE 8·3. Composition of sow colostrum and milk

	Total Solids	*Lactose*	*Fat*	*Protein*	*Ash*
Colostrum	30·0	4·5	8·5	17·0	1·0
Milk	20·0	4·5	8·5	5·5	1·0

Because the composition changes quickly after farrowing, it is important that each piglet obtains an adequate suckle as quickly as possible after farrowing in order to ensure adequate intake of colostrum.

In attempts to rear surplus piglets artificially from soon after birth, it is important that the sow-milk substitute used simulates the content of sow milk as closely as possible.

MILK YIELD

Newly-born piglets will obtain, on average, 20 ml of milk at each hourly suckling or about 500 ml in a day. Thus, to cater for a litter of ten piglets, sows should be producing five litres of milk daily at the start. Milk yield rises up to about three weeks after farrowing and then gradually declines (see Figure 8·4).

It can be seen that milk yield is much reduced by eight weeks after farrowing. When milk yield begins to decline after three weeks, piglets must be provided with supplementary food if their growth rate is to be maintained or increased. The composition and provision of creep feed is covered in the next chapter.

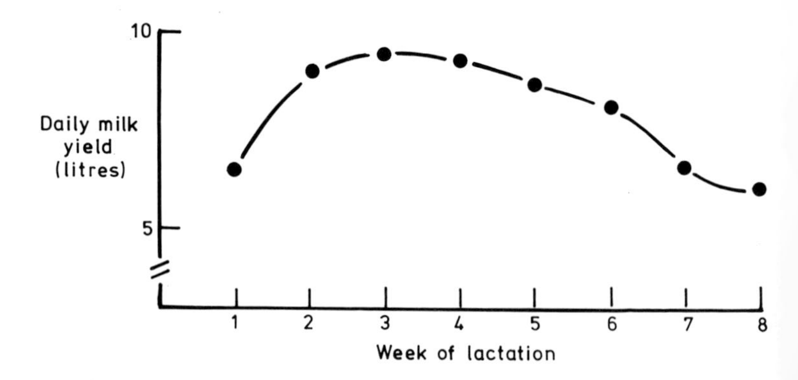

Figure 8·4. Trend in sow's milk yield.

Source: Lodge, G. A. and Lucas, I. A. M., Rowett Institute.

Since little solid food is consumed before three weeks of age, the growth of piglets up to three weeks is largely dependent on the milk yield of the dam. Thus, litter weight at three weeks of age provides the best measure in practice of the milk yield of the sow.

SUMMARY

It is obvious from the foregoing that many behavioural, biological and physical factors can adversely affect the milk production of the sow and the opportunity for each and every viable piglet born to obtain adequate nutrition from its dam. These factors must be recognised and appropriate husbandry steps taken in order to ensure optimum survival and growth in piglets. Such measures are discussed in more detail in the next chapter.

Chapter 9

MANAGEMENT OF THE LACTATING SOW AND LITTER

THE OBJECTIVE in management of the sow and litter from farrowing to weaning is to rear all viable piglets born and produce healthy, wellgrown pigs at weaning. In addition, the sow must be treated in such a way that she will come in heat and conceive promptly after weaning. The main purpose of this chapter is to deal with the welfare of the piglets so that the maximum proportion of those born are converted into sound, healthy pigs by weaning.

THE CHALLENGE TO NEWBORN PIGLETS

At birth, piglets are faced with a tremendous challenge. From the protected environment and ensured nutrition in the uterus, they have to adapt to a completely new environment and, through their own efforts, have to obtain regular and adequate nutrition from their dam in competition with litter-mates if they are to survive and thrive.

How well they meet this challenge depends on how normal and vigorous they are at birth, on the adequacy of the environment and accommodation provided for them, on the co-operation and milking ability of their dam, on the degree of competition from their litter-mates for available teats and nutrition and on the adequacy and motivation of management and stockmanship.

The factors which affect the viability, survival and growth performance of the newborn piglet must be recognised before sound husbandry systems can be evolved to optimise survival and performance.

PIGLET LOSSES

PIGLET LOSSES IN RELATION TO NUMBERS BORN

Before examining the basic causes of piglet losses, the importance of mortality must be put in perspective in relation to numbers born. Ten per cent mortality in a herd with an average of nine born per litter (leaving 8·1 reared per litter) is much more serious than 20 per cent mortality in a herd with an average of 12·5 born (leaving 10·0 reared per litter). Of course, all piglet losses are serious, but the point must be emphasised that piglet losses should always be considered in relation to the numbers present at the outset. A percentage mortality figure is relatively meaningless in practical terms unless the numbers born are specified.

DETERMINING CAUSE OF LOSS

Losses of piglets fall into two groups, namely, whole litter loss and the chronic loss of a few piglets from each litter. This chronic loss is much more serious in terms of total piglets lost. While findings from a post-mortem examination can, in many cases, provide a fairly accurate indication of causes of whole litter loss, further information is required to determine causes of death in chronic cases where one or two piglets are being lost per litter. This is because, in these cases, the factors predisposing to death are not usually detectable from post-mortem findings alone. These predisposing factors to death may involve such aspects as inadequacies in the climatic environment provided, poor farrowing facilities, ill health in the dam and excessive competition for teats and nutrition. Thus, in order to isolate these factors predisposing to chronic losses of a few piglets per litter, information on sow and piglet health and behaviour and on the climatic and physical environment provided is required. It is necessary to combine such 'case history' information with findings from a post-mortem examination in order to determine cause of death in a reasonably accurate manner.

TIMING OF DEATHS

Most piglet deaths take place in the first few days of life as indicated in Figure 9·1.

It can be seen that over 50 per cent of deaths take place before piglets are two days old. In fact, the figures underestimate the importance of this very early period with regard to the damage done. The majority of deaths occurring after the first 24 hours can be traced to happenings in the first crucial hours of life. Such factors as varying degrees of malnutrition set in from the moment of birth with piglets which are going to die. Malnutrition starting at birth does not result in sudden deaths but in a gradual loss of condition resulting in death 2–21 days after birth. The time of ultimate death is not important. What is important is the time at which the piglet was first weakened by malnutrition, chilling, injury, ill health or any other factor. Since most adverse factors affect the viability of the piglet from very soon after birth, this very early period in the life of the piglet is the one which demands attention in efforts to reduce piglet losses. This is the period when a piglet can either become firmly established or else start on the slippery slope which sooner or later leads to death.

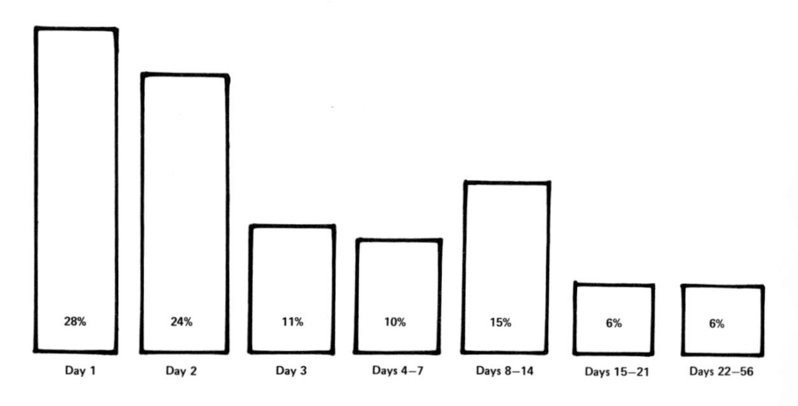

Figure 9·1. Proportion of pigs dying in different stages before weaning.

PIGLET LOSSES–THE PREDISPOSING FACTORS

Piglets vary in viability at birth because of inherent differences present before birth. Some may have genetic/developmental defects such as blind gut (atresia ani), splayleg and congenital tremor (see Chapter 3). They also vary in weight which is likely to indicate degree of maturity, energy reserves available and hence, vigour at birth.

These inherent differences in vigour and viability can be further influenced by adverse happenings such as injury and suffocation suffered during the birth process. Suffocation can result in death during farrowing (intrapartum stillbirths), or in weakened piglets at birth. These problems associated with the birth process have been covered in Chapter 6.

At birth, newborn piglets are faced with the challenge of obtaining adequate and regular nutrition from their dam. In their efforts to obtain nutrition, they have to compete with litter-mates and this competition, while being fairly minimal in small litters, can be extremely severe in large litters, especially when there are insufficient functional teats exposed to provide for all piglets in the litter. Obviously, pigs with any defects, and the smaller, less vigorous pigs, are likely to suffer most in such a fiercely competitive situation. The factors influencing the chances of each piglet obtaining an adequate and regular suckle have been covered in Chapter 8.

The physical and climatic environment provided for farrowing can have a major influence on piglet deaths and the important provisions in this regard were dealt with in Chapter 7. In particular, the need for a cosy, well-protected environment for piglets and of farrowing pen arrangements (type of crate, floor and situation and adequacy of piglet creeps) designed to minimise risk of crushing were emphasised.

Chapters 6, 7 and 8 therefore serve as important background to the present chapter.

The extent and causes of piglet losses will vary from farm to farm according to such factors as numbers born, birthweights, soundness and health of stock and adequacy of the farrowing accommodation.

On one unit examined, the extent and causes of deaths were found to be as in Tables 9·1 and 9·2.

TABLE 9·1. Piglets per litter

	Number
Mummified*	0·5
Stillborn: Prepartum deaths**	0·1
Intrapartum deaths**	0·4
Born alive	10·8
Deaths to weaning	2·0
Weaned	8·8

* Mummified = Partially reabsorbed piglets, some of which were only 20 g at farrowing.

** Stillborn. Prepartum = dead before farrowing. Intrapartum = dying during farrowing.

TABLE 9·2. Causes of death in liveborn piglets

Cause of death	*Mean birthweight (kg)*	*Number per litter*	*Per cent of total*
Congenital/genetic abnormalities	1·13	0·2	12
Extremely weak at birth (unable to move)	0·80	0·2	9
Weakness caused by partial suffocation during farrowing	0·81	0·1	6
Crushing (normal thriving piglets)	1·15	0·4	18
Starvation (normal piglets at birth)	0·91	0·9	43
Primary infection	1·21	0·1	6
Miscellaneous (savaged, born alive but suffocated in afterbirth, etc.)	1·21	0·1	6
	TOTALS	2·0	100

The average birthweight of stillborn piglets was 0·96 kg, of all live-born piglets 1·24 kg, of liveborn piglets dying before weaning 1·0 kg, and of those surviving piglets 1·32 kg.

The specific figures in Tables 9·1 and 9·2 may or may not have any relevance to other units. They are presented primarily to indicate the major categories of piglet loss. The extent of losses and the relative importance of each cause of loss will differ from unit to unit.

However, it is noticeable that it tends to be the lighter piglets which die and this is a universal finding.

It is important to discuss these separate sources of loss in more detail, to isolate the causes for loss and to suggest methods for reducing losses to a minimum.

RELATIONSHIP OF BIRTHWEIGHT TO MORTALITY

It is well established that low birthweight piglets are at a disadvantage relative to larger piglets because of their greater proportional surface area relative to body weight and thus their greater liability to lose heat and die from chilling. Smaller piglets at birth are also likely to have lower energy reserves at birth which will place them at a further disadvantage in sub-optimal climatic conditions.

Consequently, piglets of low birthweight have an inherent disadvantage relative to those of high birthweight. This disadvantage is, however, greatly accentuated if such piglets have litter-mates which are larger than themselves since they are at a distinct physical disadvantage in the intense competition for suckling places at the udder. The hourly milk 'let down' lasts for only some 20 seconds and low birthweight piglets can experience great difficulty in fighting their way to the udder as the litter is 'grunted up' by the sow, in succeeding in getting into position to grip a teat quickly enough and maintaining this position against physical superiors during the very short time the milk is actually available.

Thus, all studies have demonstrated the fact that lighter piglets at birth suffer higher mortality (see Figure 11·2, Chapter 11) and some have demonstrated the fact that the greater the variability in birthweight, the higher the mortality. The importance of variation in birthweight in relation to mortality is demonstrated in Table 9·3.

Thus, litters in the two categories, i.e. those with uniform and those with variable birthweights were almost identical in terms of numbers born and mean birthweight. However, fully one pig more was lost from the variable litters.

What this information tells us is that a pig of low birthweight has a very low chance of survival among much larger litter-mates but has a considerably improved chance of survival among piglets of its own size.

TABLE 9·3. Mortality in litters with uniform and variable birthweights

	Uniformity of birthweights within litters	
	Uniform litters	*Variable litters*
Number litters	32	32
Mean birthweight (kg)	1·2	1·2
Number born alive	11·3	11·4
Losses	2·0	3·1
Weaned	9·3	8·3
Per cent mortality	17·7	27·2

Uneven piglets in the litter means that there is unfair competition for teats at the hourly feeds with the small ones being deprived by their physical superiors in the scramble to get to the udder, to find a teat and to grip it properly so as to get an adequate suckle during the very short time that milk 'let-down' lasts. Consequently there is a much higher mortality among smaller piglets as a result of malnutrition. Therefore in attempting to improve the survival chances of pigs of lower birthweight it is important that their litter-mates be of much the same size.

Uniformity in birthweights within litters cannot be achieved by manipulating feeding or by selection. There is some indication that crossbred litters are more uniform than purebred litters, while older sows tend to have less uniform litters. However, the most effective way to achieve increased uniformity in birthweight within litters is to batch-farrow and cross-foster piglets between simultaneously farrowed litters so that all the small piglets would be allocated to one sow and the heavier ones to the other.

Evening up birthweights within litters by cross-fostering piglets between sets of two sows farrowing within six hours of each other produced the results relative to controls shown in Table 9·4.

In this work, cross-fostering to even up birthweights within litters brought about a useful improvement in survival. The bigger the litter size and the more variable the birthweights, the more useful such cross-fostering is likely to be in improving survival. It is important that such cross-fostering takes place within six hours or so of farrowing for maximum benefit to accrue.

TABLE 9·4. Effects of cross-fostering so as to equalise birthweights within litters

	Control litters	*Cross-fostered litters*
Number litters	18	18
Number piglets:		
born alive	10·9	11·0
weaned	9·5	10·2
Weights (kg):		
birth	1·3	1·4
6 weeks	10·5	10·8
Per cent mortality	12·8	7·6

REDUCING LOSSES

Stillbirths

Stillbirths constitute a serious source of loss in the pig herd, averaging half a pig per litter or one pig in every second litter. Such losses approach one million piglets per year in the UK. On average, in a 100-sow herd, over 100 piglets per year will be stillborn. Most of these—about 80 per cent—are intrapartum stillbirths, i.e. they would have been alive at the start of farrowing but would have died before birth because of suffocation following premature breaking of the umbilical cord, detachment of the placenta or other factors associated with prolonged delivery. The fact that most stillbirths are occurring so soon before farrowing increases the chances that such losses can be reduced.

If there is a high incidence of stillbirths in a herd, it may be worth checking the haemoglobin level in the sows. Normal levels are between 12 and 16 mg per 100 ml blood and low levels have been associated with a higher incidence of intrapartum stillbirths. However, where normal diets are fed, low haemoglobin levels in sows are very unlikely.

A higher incidence of intrapartum stillbirths is experienced in bigger litters and in older sows (see Figure 9·2), probably because farrowing will tend to be prolonged in such cases.

Since intrapartum deaths increase in older sows, attention must be paid to timely culling of sows and, after considering trends in stillbirths, numbers born alive, birthweights and numbers reared, some producers cull sows routinely after the sixth litter. An older sow which produces, say five livebirths and seven stillbirths, has been kept for one litter too many.

Most intrapartum stillbirths occur towards the end of farrowing. This is often due to exhaustion (secondary inertia) and is compounded in old sows by loss of muscular tone in uterine muscles. As outlined in Chapter 6, the drug neostigmine which stimulates smooth muscle contractions, gives promise of assisting the safe delivery of later born piglets, particularly in old sows. In experimental work at Aberdeen where neostigmine has been administered to sows following the birth of the fourth or fifth piglet, useful reductions in intervals between births and in the incidence of intrapartum stillbirths have been achieved. In the future, the use of drugs such as neostigmine in this way could become an important component of farrowing management systems following induction of farrowing using analogues of prostaglandin F 2α.

Part of the constant routine to minimise the incidence of intrapartum stillbirths should be the regular monitoring of the farrowing process followed by provision of timely and appropriate assistance as outlined in Chapter 6.

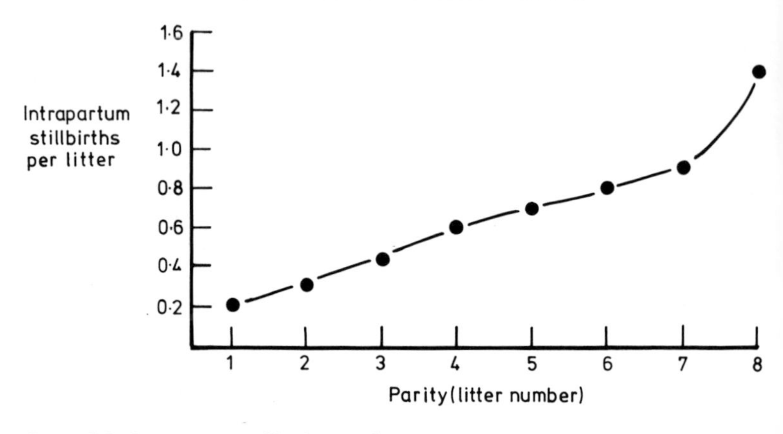

Figure 9·2. Intrapartum stillbirths in relation to parity or litter number.

Genetic and developmental defects

Among the problems encountered are blind gut (or atresia ani), congenital splayleg, trembles and heart defects. Blind gut is usually fatal in the male, although sometimes the rectal passage can be continued to the exterior by a simple operation. Further attention may be required, however, to ensure the passage remains open. Piglets with splayleg may often recover fully if they can obtain adequate and regular feeding. Thus, some assistance given to these piglets to suckle initially may be all that is required to give them the impetus and confidence to fend for themselves thereafter.

The control and prevention of these problems have been covered in Chapter 3.

Ill health

Injuries, an adverse climatic environment and malnutrition can all predispose to ill health and therefore every attempt should be made to eliminate these factors which lead to disease.

Having a healthy herd provides an essential foundation to an efficient unit with minimal disease losses. Those producers who keep closed herds appear to enjoy many advantages. If this is not feasible, then replacement stock should be obtained from one reputable source. Dividing farrowing and rearing units into sections so that an 'all-in, all-out' policy can be followed, coupled with regular cleaning and disinfection between batches, and good hygiene in general, all help to produce the healthy pigs which give us a head start in attempts to minimise piglet losses.

For further reading on these aspects consult Chapters 3 and 4.

Crushing and malnutrition—as affected by farrowing arrangements
By far the most important factors in piglet losses are crushing and malnutrition (or starvation). These two factors together account for between 50 and 80 per cent of piglet deaths. The relative importance of these two factors will vary with the particular circumstances prevailing on any pig unit. The incidence of crushing will be high in those herds in which farrowing facilities are poor, piglet creeps are cold or badly situated, sows are restless, old and clumsy, and piglet birthweight is low.

Starvation will be a serious cause of piglet losses in herds where there are large average numbers born, birthweights are low and variable, number of functional and exposed teats per sow are reduced, incidence of post-farrowing agalactia (milk shortage) is high, climatic environment is poor, and batch farrowing is not practised.

However, crushing and malnutrition are often connected since an injured piglet is less capable of succeeding in getting a decent regular suckle. On the other hand, a piglet failing to get a regular suckle is weakened as a result, and is more liable to be crushed. Often the same factors predispose to crushing and malnutrition. One such factor is an adverse farrowing house environment.

Newborn piglets in a litter require a temperature of between 28 and 30°C for comfort and optimal performance. If piglet creeps are cold, piglets will tend to lie against the sow for warmth, thus leaving themselves more liable to crushing. In such an environment, piglets will be wasting much of the energy from milk just to keep up their body temperatures, and therefore less nutrients will be available for growth and to keep them thriving.

Poorly designed farrowing pens may also result in high losses from crushing and/or malnutrition for several other reasons:

- The farrowing crate may allow the sow too much latitude and therefore she will not be able to exercise sufficient control when lying down.
- The floor surface may provide an unsafe footing for the sow or may affect adversely the mobility, and therefore escape possibilities, of the piglets.
- The pen floor may be uncomfortable for the sow and/or the piglets because of wetness, an abrasive surface or sharp edges on slatted floors.
- The piglet creep areas may be too far away from the udder, they may be too small or too cold and all these situations will tend to encourage piglets to lie against the sow where they will be more prone to crushing.

- There may be too many farrowing pens in the one house and so any undue disturbance in any one pen in the house is likely to have an adverse effect on a large number of sows and litters. Houses may be too brightly lit and result in excessive restlessness of sows with consequently increased risk of crushing.
- The farrowing arrangements may effectively reduce the rearing capacity of the sow. This may arise because the floor surface is not sufficiently comfortable to encourage the sow to rotate her body fully at suckling and so expose the maximum number of teats to piglets. In addition, if the bottom bars of the farrowing crate are not placed at the correct height, they may restrict access of piglets to the upper row of teats.

Methods of avoiding the above problems and of meeting the desirable objectives in designing farrowing arrangements were covered fully in Chapter 7.

Provision of a sound farrowing facility, purpose-built to provide fully for the welfare of the sow and her newborn piglets, provides a sound baseline for organising a system to achieve high piglet survival rates.

Given viable piglets at birth which are free from genetic/developmental defects and which have a good health status, whose safety from crushing is catered for by a sound farrowing/rearing facility, the next vital step in striving for high piglet survival is to attempt to ensure adequate and regular nutrition for all viable piglets born.

Trying to ensure feed for all

Trying to ensure adequate nutrition for all viable piglets born is dependent on many factors:

Farrowing pen floor and crate

It was noted in the last section that the farrowing pen floor should be comfortable in order to encourage the sow to rotate her body at suckling so as to improve exposure of all teats on the lower row to piglets. It was also noted that the bottom bar of the crate should be adjusted so that it does not impede piglets' access to the top row of teats at suckling.

An adequate environment

Baby piglets have only small reserves of energy which they will quickly use up if the temperature is too low. This results in high mortality. Providing an adequate temperature, in effect, gives an energy boost to the piglet which helps greatly in getting it established. A newborn piglet in a litter requires a temperature of 28 to 30°C and the cheapest way of providing this is in the form of a micro-environment rather than by space heating.

Placing a cover over the heat sources helps to preserve the heat produced by piglets and heater, concentrates the heat where piglets require it and helps to reduce draughts and convection currents. Preventing draughts is very vital as a slight draught can have the same adverse effect on the piglet as a 4°C drop in temperature. Certainly placing a cover over the heat source makes inspection of piglets more difficult but a little extra trouble in inspecting piglets is likely to be a relatively small price to pay for economies in energy costs and a more comfortable environment for piglets.

Availability of milk

Milk can be unavailable to piglets because of:

- Agalactia in the sow.
- Surplus or supernumerary piglets.

Agalactia or milk shortage can be caused by various factors (see Chapter 8). The most useful step the stockman can take to minimise this condition is to keep a constant check on the well-being of sow and piglets after farrowing. The appetite and rectal temperature of the sow and restless activity and loss of condition in piglets are the best indicators of the condition. It is important to spot the condition quickly and arrange for prompt veterinary treatment as piglets cannot stand milk deprivation or shortage for any length of time because of their low energy reserves.

While sows are suffering from agalactia, piglets should be given limited amounts (by no means *ad libitum*) of milk substitute to tide them over until the sow is milking well again.

Piglets may be deprived of milk because there are not enough functional teats to go round. Many sows, especially older ones with more pendulous udders, fail to expose some of the more hindmost teats on the lower row at suckling and this effectively reduces their rearing capacity. It is important to spot this problem promptly after birth and foster surplus piglets within two or three days of farrowing to newly farrowed sows with spare rearing capacity.

Fostering

Among the sow's shortcomings are her variable and unpredictable litter size and rearing capacity and variation in birthweight within the litter. This results in some sows having too many and others having too few piglets in relation to their rearing capacity. Fostering surplus piglets to sows farrowing at the same time with spare rearing capacity is the obvious solution. Fostering is a very straightforward technique in pigs and all stockpersons exploit it although few exploit it as fully as they might usefully do. Obviously, effective batch farrowing would increase the scope for fostering and this is to be discussed shortly.

In planning fostering policy, one must first assess the 'rearing capacity' of each sow. 'Rearing capacity' is best defined as the number of functional teats which a sow is exposing at suckling to her piglets. Piglets in excess of the 'rearing capacity' of a specific sow should obviously be fostered to a simultaneously farrowed sow having spare 'rearing capacity'. Normally, surplus piglets may be fostered on to a sow with spare rearing capacity which has farrowed up to three days previously. After this time, spare glands tend to dry up.

Uneven piglets in the litter means that there is unfair competition for teats at the hourly feeds with the small ones losing out and suffering high mortality as a result of malnutrition. Cross-fostering piglets between litters which have farrowed within about six hours of each other can help to solve this problem, i.e. the two litters can be grouped together, the heavier piglets placed on one sow and the lighter ones on the other. Such a practice has produced useful dividends in improving piglet survival (see Table 9·4). Crossfostering is likely to be most useful in herds with high average numbers born and with a lot of variation in birthweight.

A small piglet has a relatively poor chance of survival when most of its litter-mates are much larger but such a small piglet can survive perfectly well when its litter-mates are much the same size as itself.

ASSISTING WEAKLY PIGLETS

As has been indicated above, one way of boosting the survival chances of lower birthweight piglets is to arrange for them to be members of litters all of about the same size by the process of cross-fostering. This helps such piglets to be more competitive in striving to get a teat and an adequate suckle regularly.

The smaller weaker piglet can also be helped by ensuring that it gets to the heat source as soon after birth as possible and by assisting it to get to the udder and to suckle if it is experiencing difficulty in doing this itself. It is amazing to observe how quickly the attitude of such a weakly piglet can change after it has been assisted to suckle successfully. Previously it can have been totally disinterested and misdirected in its attempts to obtain life-giving nutrition. After it is held to a teat with a plentiful supply of colostrum and some colostrum 'milked' into its mouth initially, it can quickly get the idea and there is a sudden amazing change in its behaviour and attitude. One can feel its whole body flexing in its determination to feed and to live, whereas previously it seemed determined to die.

Other techniques for assisting such weakly piglets include giving sow colostrum in fairly small quantities (10 to 20 ml at a time) by syringe over the tongue or by stomach tube (cow colostrum is a possible alternative as this has useful protective and nutritional properties for piglets) and intraperitoneal injections (5 to 10 ml) of a 20 per cent solution of glucose and water at blood temperature. It is important to avoid over-dosing weakly piglets with colostrum or glucose because this could induce scouring and reduce the drive to suckle. After providing weakly piglets with limited amounts of these supplements, regular attempts should be made to direct them to the teats and assist them to suckle.

Some may question the worthwhileness of saving these small weakly pigs. When a good environment and farrowing arrangements are provided, all the piglet requires is that initial assistance to get it established and thereafter it can fend for itself. Once it gets established it has the ability thereafter to grow and convert feed as efficiently as its larger litter-mates.

BATCH FARROWING

Batch farrowing is very important for reducing piglet losses from malnutrition because it increases opportunities for fostering. Batch farrowing is automatic in very large herds because several farrowings will be taking place at a time. In smaller herds one can only resort to batch weaning in striving for batch farrowing but, with variations in the weaning-to-service interval and in the gestation period, there can often be disappointments about the degree of batching of resultant farrowings. The advent of analogues of prostaglandin F_{2a} is likely to prove useful for synchronising farrowings more effectively.

SYNCHRONISED FARROWING

It is clear that maximum piglet survival depends on applying in practice a multiplicity of important principles and measures. Since a proportion of piglet deaths occur or can be triggered off by events in the first few hours of life, regular supervision during and for a few hours after farrowing is likely to be very beneficial for piglet survival. However, without synchronised farrowing, such supervision may not be cost effective or socially acceptable on many units.

Thus, the advantages of supervision at farrowing are recognised but such supervision is rare because of the unsocial hours involved and because of doubts on cost effectiveness.

The advent of substances such as prostaglandin analogues to effect synchronisation of farrowings gives promise of providing the necessary incentive for pig producers almost to revolutionise their farrowing management. These drugs will help them apply the principles and measures which were known to be sound but which were not fully applied because timing of farrowings were relatively unpredictable and because they could be taking place in any of the 24 hours of any day and in any of the seven days in any one week.

With distinct possibilities now of achieving good batch farrowing using prostaglandin analogues and confining the majority of farrowings to normal working hours on perhaps three days in each week, a farrowing system can be evolved with the objective of minimising piglet losses. Such a system might involve some or all of the following measures:

- Administration of a substance such as neostigmine after the birth of four or five piglets to stimulate contraction of uterine muscles with the objective of reducing incidence of intrapartum stillbirths and of later-born livebirths suffering adversely from suffocation during delivery.
- Revival of piglets which are very anoxic at birth.
- Removal of membranes from any piglets enveloped in such at birth.
- Tying cords of navel bleeders and administration of iron dextran (e.g. $\frac{1}{2}$ cc) to piglets with very pale appearance at birth.
- Assisting very weakly piglets by administering a solution of glucose or of sow or cow colostrum by syringe or stomach tube.
- Placing of weakly piglets under the heat source and assisting them to find a teat and to suckle.
- Enclosing piglets within safety of creep area when their dams are restless and when piglets therefore are in danger of being crushed.
- Monitoring piglet delivery times and arranging assistance for the sow when her continued contractions and a long interval from previous birth (e.g. in excess of 30 to 40 minutes) indicated dystocia.

- Monitoring rectal temperature, sow appetite, sow milk 'let down' and piglet contentment as criteria of conditions leading to agalactia and arrangement of prompt remedial treatment.
- Assessment of 'rearing capacity' of each sow by assessing number of functional teats exposed to piglets at suckling and matching litter size to such rearing capacity by fostering on or off as required.
- Cross-fostering to even up birthweights within litters.
- Taking appropriate measures to prevent savaging of piglets.

Fairly extensive studies on a range of institutional and commercial pig units are now in progress to assess the effectiveness of prostaglandin analogues in synchronising farrowing and in reducing piglet losses. Results in terms of effective synchronisation of farrowing are promising as indicated in Tables 9·5 and 9·6.

TABLE 9·5. Plan of induced farrowings using injection of Prostaglandin F_2a analogue, CLOPROSTENOL

Injection time: Thursday 9 am

Sows injected	45 due to farrow (115 days) on Friday
	41 due to farrow (115 days) on Saturday
	9 due to farrow (115 days) on Sunday
	—
Total sows treated =	95

TABLE 9·6. Timing of induced farrowings on Friday

	Number	*Per cent*
Farrowing before 7.00 hr	25	26
Farrowing between 7.00 and 17.00 hr	67	71
Farrowing between 17.00 and 19.00 hr	3	3
Farrowing after 19.00 hr	0	0

The aim of the induction programme detailed in Table 9·5 was to get the maximum number of sows farrowing during normal working hours (7.00 to 17.00 hr) on the day following injection. While 26 per cent of sows farrowed before 7.00 hr, only 3 per cent farrowed after 17.00 hr leaving 71 per cent farrowed within the working day.

While induction and synchronisation of farrowings have been effective, when farrowings are induced too early, piglet viability at birth is reduced. The relative safety of inducing one, two and three days before the due-to-farrow date relative to inducing sows to farrow on the due date may well differ according to the quality of the farrowing arrangements. Inducing farrowings before term or due date means slightly smaller and less mature piglets at birth. If farrowing arrangements are good, a slight reduction in maturity and weight at birth may have little, if any, adverse effect. However, smaller, less mature piglets are likely to suffer more in a poorer farrowing environment.

With uncertainties of this nature still existing, it is not yet possible to make a recommendation on how much earlier farrowings may be induced in different situations without having an adverse effect on piglet viability at birth. However, for many situations in practice, it is unwise to induce farrowings more than one day prior to the 'due-to-farrow' date.

COMPLEMENTRY REARING

On liquid feed

'Complementary rearing' is the term given to the artificial rearing of some piglets from a litter, while the remainder are reared naturally by the sow. In minimising deaths from malnutrition, much emphasis was placed on batch farrowing and fostering surplus piglets from large litters. The anomaly of this situation is that if one has supernumerary piglets to provide for in this way, one is dependent on the farrowing of some small litters to provide the necessary spare 'rearing capacity'. Sometimes, fortunately or otherwise (depending on how one looks at it), such small litters do not materialise and this results in viable piglets which are surplus to batch or herd rearing capacity. This situation creates a demand for a suitable system for artificial rearing of such surplus piglets.

If in a batch of farrowing sows one dies or suffers from agalactia, the surplus piglet problem will be accentuated and the need for artificial rearing further increased. In addition, in herds with high average numbers born, there are often one or two small piglets in a litter which are perfectly viable but which find it impossible to compete effectively with their bigger fellows and these under-privileged pigs are further candidates for an artificial rearing unit if such was available.

The essential features of such a system are:

- It should be capable of coping with piglets from about six hours of age. Piglets should be left on the sow for this time in an attempt to ensure intake of some colostrum to provide protection against prevailing infection.
- Piglets should be housed singly or in small groups in a very

comfortable micro-environment, free from draughts (temperature around 35°C at the start).
- Feeding may be from teats or bowls. It is easier to train piglets to use bowls rather than teats.
- Feeding should be on a 'little and often' basis. If feeding is automatic, then simulating the sow by feeding once per hour with about 20 cc offered at each feed is to be recommended. If hand feeding is practised, then feeding four to six times daily, with about 100 cc offered at each feed, is advocated.
- Strict hygiene should be observed and feeding utensils should be cleaned regularly.
- A suitable diet is cow colostrum which has useful health protective qualities for piglets and at the same time is an excellent nutrient. Various sow-milk substitutes are also available.
- In the event of scouring, the level of food intake should be decreased. Overfeeding is one sure way of inducing scouring as this results in an overloading of the digestive system.
- Piglets can be transferred to solid feeding (milk substitute) at about 3 kg liveweight.

Some producers with high average numbers born are able to rear 0·5 to one piglet per litter artificially so that this is a useful addition to those reared by the sow. Conversion of milk solids to liveweight gain in this period is excellent—1 kg of liveweight gain from 1 kg or less of milk solids.

While some energetic and progressive stockpersons successfully operate hand-feeding systems for artificial rearing of surplus piglets, there is a need for a fully automatic system in terms of feeding and regular cleaning and disinfection of the feeding utensils in order to reduce the chores of feeding and cleaning out.

On solid feed

In herds practising five- to eight-week weaning with large numbers born per litter and high initial survival, it is often found that some piglets in such large litters thrive well at first but around 10 to 15 days of age they then start to lose condition and their litter-mates grow away from them. If this situation is neglected, such piglets either die before weaning or else end up as runts and problems at weaning. The main reason for one or two piglets in large litters losing condition in this way is that they are being deprived of a regular and adequate suckle. This is because either their teats have dried up or else they have physical difficulty in getting into the udder as the litter grows and the space at the udder remains the same.

Plate No. **33**

Often in large litters, most piglets thrive reasonably well but one or two may be in danger of becoming 'nutritional runts' unless alternative provision is made for them.

The main cause for such ill-thriving being nutritional, the solution lies in catering for these piglets by fostering them on to a sow with spare capacity or else, if a large number of such piglets exist, it is worthwhile considering setting up a small early-weaning unit in a corner of the farrowing/rearing house or in a separate room. This could consist of a flat-deck cage unit and there must be an adequate controlled environment. The ill-thriving piglets (not the whole litter) can be transferred to such a unit at around 10 to 15 days of age (and not less than 3 kg liveweight) before they lose too much condition and given limited quantities of an early-weaning diet in dry form. This diet should be milk based initially and this can be changed gradually to a largely cereal-based diet around 6 kg liveweight. Artificial rearing of these potential nutritional runts works well provided the appropriate environment and nutrition are provided. By 50 days of age these piglets considerably narrow the gap on, or even overtake, their sow-reared litter mates. A typical example is shown in Figure 9·3.

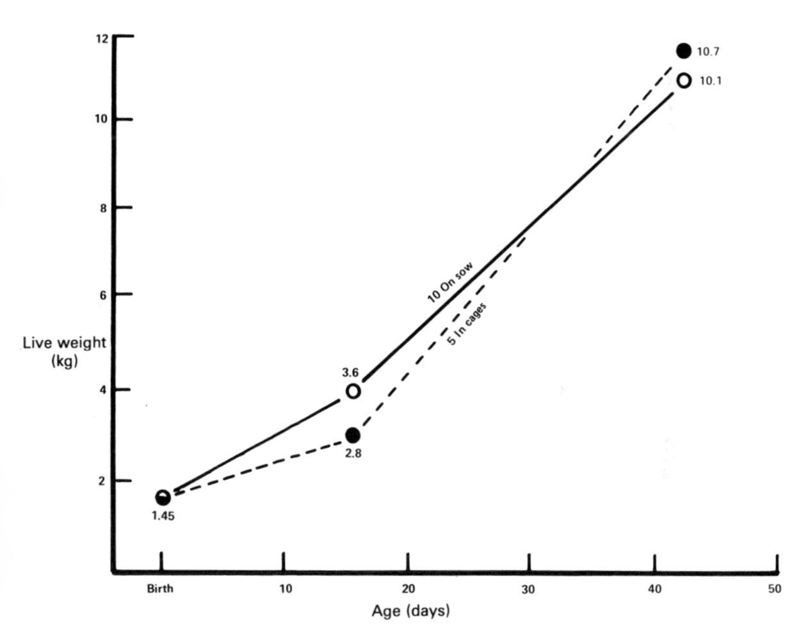

Figure 9·3. A typical example of complementary rearing. Relative growth of five early-weaned 'poorer doers' weaned at 16 days and ten sow-reared litter-mates from a litter of 15.

Such a practice on one farm has produced the results shown in Table 9·7. In a sample of ten litters, an average of ten piglets were left on the sow and the poorest seventeen piglets were early weaned.

The feed cost of rearing these piglets was therefore £0·35 per kg at a time when one kilo of weaner liveweight was worth about £1.05.

In this case, the sows reared an average of ten piglets, while an extra 1·7 piglets per sow were reared in cages, making an output per sow of 11·7 piglets. Thus, the system can be usefully used in herds weaning at five to eight weeks of age to rear the less well thriving piglets from large litters from around 10 to 15 days of age and provide a useful boost to sow output. It is also useful, of course, for rearing very small litters from the same stage and allows earlier culling or rebreeding of the sow. When a sow dies one to three weeks after farrowing, orphan pigs can also be catered for.

TABLE 9·7. Performance of 17 cage-reared piglets from 10 litters

Number of pigs	17
Age at start (days)	16
Weight at start (kg)	2·5
Weight at 6 weeks (kg)	10·2
Mortality	0
Food consumed (kg) milk pellets	0·95
cereal pellets	3·0
weaner meal	9·1
Food conversion ratio	1·77
Liveweight gain per day (kg)	0·22
Food cost per kg liveweight gain*	£0·35

* At May 1982 prices.

The main essentials of such a system are as follows:

- Minimum weight at transfer to dry food, about 3 kg (7 lb).
- Transfer promptly—don't wait until piglets lose too much condition.
- Provide a good uniform micro-environment (21 to 25°C without draughts).
- Start off on milk-based pellets (21–24% CP)—limit quantity fed to prevent scour—keep edge on appetite.
- At 5–6 kg liveweight transfer gradually to cereal-based pellets (18–20% CP).
- At 12–15 kg liveweight transfer to weaner meal and move out of cages into suitable follow-on pen.
- Around 25 kg can mix with rest of weaners.

An alternative way of dealing with these piglets is to gather them together and foster them on to a newly-weaned sow which is still milking reasonably well. This has its advantages in that no further capital investment is required and the sow rather than the stockperson is saddled with the work of rearing.

On the other hand, the disadvantages of being dependent on such a sow are that one may not be available just when required and good milking sows on five- to eight-week weaning systems may already have been brought well down in condition.

REDUCING PIGLET LOSSES—A SUMMARY

The main essentials in achieving higher piglet survival and more weaners per litter can be summarised as follows:

1. Good health.
2. Efficient breeding stock (as free as possible from genetic defects).
3. Gilts with good underlines (at least fourteen well-developed, well-spaced teats).
4. Good management in pregnancy.
5. Crossbred sows and litters.
6. Good farrowing pen and crate design (top priority to welfare of piglets and sow).
7. Adequate micro-environment for the baby pig, especially temperature.
8. Batch farrow.
9. Availability of knowledgeable, skilled and motivated stock persons.
10. Stockpersons given enough time to exercise skills.
11. Feed suckling sows for adequate milk production.
12. Provide particular attention at farrowing to reduce losses due to intrapartum stillbirths and to attend to savaging, restless sows and piglets born in afterbirth. Assist weakly piglets by placing them in a warm creep, and directing them to a teat and assisting to suckle. Very weak piglets may be given glucose and/or sow or cow colostrum.
13. Regular checks on sow health (appetite and rectal temperature combined with behaviour and appearance of piglets). Have treatment administered promptly.
14. Assess rearing capacity of sow (milk availability, functional teats exposed etc). If inadequate, foster surplus piglets.
15. Cross-foster to even up birthweights within litters.
16. Prevent anaemia.
17. Make provision for potential nuritional runts, e.g. fostering or complementary rearing.
18. Adequate supplementary feeding.
19. Cull poor and older sows rigorously.

These nineteen factors can be regarded as essential components or building bricks of a piglet production unit incorporating the sow, the accommodation, the stockman, the system of operation and all other inputs. We can look upon these building bricks as forming the foundation and corner stones of a structure supporting a roof on which is perched a litter of piglets, some rather precariously, as in Figure 9·4. If vital foundation and/or corner stones are missing, the roof will tilt and piglets closest to the edge will fall to their destruction. However, the more complete the foundation and the superstructure are, the more capable will the system be of supporting all members of the litter.

There is no single magical secret for maximising piglet survival. Survival is dependent on many foundation and corner stones and only if all of these are provided and cemented into the system can we hope to achieve high survival rates in piglets. If we try to scrimp on the foundation or the superstructure, then the result will be inevitable; the roof will tilt and those piglets in the most insecure positions will be lost. The greater the number of vital building bricks which are inadequately provided in the system, the greater will be the tilt on the roof and the greater the number in the litter which will fall to perdition.

SURVIVAL TARGET

A survival target of 100 per cent is unrealistic. It is difficult to justify much effort in trying to save piglets which are abnormal at birth, such as those with serious developmental or genetic defects, e.g. those serious cases of blind gut and congenital splayleg and those piglets which are extremely weak at birth because of injury or oxygen starvation during birth due to prolonged or difficult farrowing. This is not to say, of course, that we should neglect the source of these losses. However, these problems fortunately are of fairly low incidence and, in most herds, less than 4 per cent of piglets are abnormal or defective at birth. For practical purposes we can write this 4 per cent off in the short term.

This leaves 96 per cent of livebirths which are normal and viable at birth. Therefore, a very high proportion of piglets born in the great majority of herds have the ability to live and thrive if given adequate conditions. The reason why mortality levels average 13 per cent on pig units in the UK, and not 4 per cent, is not primarily due to defects in the piglets at birth but is because of the inadequate facilities and management provided for piglets.

Of course, efforts to reduce piglet losses must be cost effective in order to make it worthwhile. The reasons for generally high levels of piglet losses will vary from unit to unit and the cost effectiveness of efforts to reduce losses must be considered in the light of circumstances on each individual unit. On some units, the owner, or more especially, the stockman, may look upon a 15 to 20 per cent piglet loss level as being normal. When such an attitude is ingrained, one is usually knocking one's head against a brick wall in making recommendations for reducing losses.

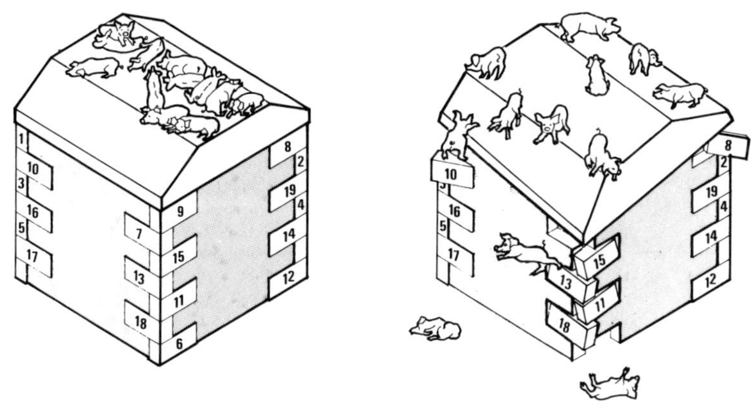

Figure 9·4. An analogy of a piglet rearing system:
1. A sound system with all vital components provided and all piglets firmly established.
2. An unsound system with one or two vital components missing or inadequately provided. This puts more piglets at risk and increases piglet losses.

When high losses are being experienced and the major contributory factor appears to be some aspect of basic design in the farrowing arrangements, then it may be more difficult and costly to rectify the problem.

The situation in which it will be simplest to achieve high survival figures is in planning a new unit. With the maternity arrangements in such a unit planned with top priority given to the welfare of the new-born piglet and the sow, careful synthesis of a sound farrowing system involving batch farrowing and sow and piglet care, and employment of knowledgeable, dedicated and motivated stockmen, high survival rates can be achieved.

There are now emerging a handful of units in which the recommendations for reducing losses listed above are being implemented by skilled stockmen. Number born alive per litter is about eleven and piglet survival rate is 95 per cent. This is illustrated in Table 9·8. It can be seen that four herds out of this sample of 305 herds are achieving a survival rate in excess of 95 per cent. 63 herds (21 per cent of total) are achieving a survival rate of between 90 and 95 per cent. With over eleven livebirths per litter, herds with a survival rate in excess of 90 per cent are weaning over ten piglets per litter, on average. This is a very creditable achievement. The stockmen operating these units, in which the whole farrowing management system has been well thought out, have the will to achieve a high level of efficiency and are given enough time in their working day to pay the necessary attention to all the small details.

TABLE 9·8. Preweaning mortality of livebirths in herds averaging over eleven liveborn pigs per litter

Mortality of livebirths (per cent)	*Number of herds*	*Proportion of herds*
0–5	4	1
5–10	63	21
10–15	143	47
15–20	77	25
20+	18	6
	305	100

Source: *MLC Commercial Pig Production Yearbook* (1980).

The old Scottish saying, "Mony a mickle maks a muckle" (put all the little bits and pieces together and you end up with quite a sizeable product), sums up aptly the sow and litter care needed to achieve high survival rates in piglets.

ROUTINE TREATMENT OF PIGLETS: IDENTIFICATION

Careful recording of the pig herd in both physical and financial terms and regular analysis of these records is absolutely essential in modern pig production for the prompt detection and remedying of problems. Animal identification is an essential basis of recording and piglets may be given either an individual or litter number. Piglets may be identified at birth or at any time up to weaning, the most common methods being ear notching and tattooing.

CLIPPING TEETH AND TAILS

The 'eye' or canine teeth can be needle sharp at birth and, as the piglets fight with each other at the udder in settling the 'teat order' both the piglets and the udder can be severely scratched. These wounds may allow entry of secondary infection and can cause considerable upset to the sow, with consequent risk of overlying and possibly, savaging of piglets. Thus, routine clipping of the upper and lower canine teeth is to be recommended.

Clipping off the end of the tail at birth, leaving about 2 cm on the piglet, is a simple operation and a useful deterrent to tail biting in the finishing pens.

WORMING

(See Chapter 3.)

CASTRATION

When the necessary care is exercised, performance is not adversely affected whether castration takes place at one or eight weeks of age or at any time in between. It is a simpler operation to castrate at an early age, e.g. at one week rather than later on.

Strict hygiene should be exercised with regard to instruments and personnel and it is important that the incision be of sufficient length to allow adequate drainage and that the cord should be pulled and not cut. (But see Chapter 16, 'Welfare'.)

ANAEMIA PREVENTION

Newborn piglets have only a limited store of iron. Milk is low in iron and unless pigs have free access to iron (as in outdoor situations) anaemia will develop in two to three weeks. In this case the anaemia is due to failure of haemoglobin production—a constituent of the red blood cell. Pigs with free access to iron only absorb enough to 'top up' the small stores that are necessary in haemoglobin synthesis.

Clinical signs of deficiency

The onset of clinical signs will depend on the presence or absence of trigger factors, the cleanliness of the pen and rate of growth of the piglet. Affected piglets usually become anaemic, look pale, listless and lethargic around 14 to 21 days of age. A pale yellow scour often accompanies these symptoms. Death usually occurs between four to six weeks of age in untreated piglets.

Factors affecting severity of disease

Cold draughty pens, intercurrent disease (infections, parasites) and haemorrhagic conditions will increase the requirement for iron, while dirty pens will provide more iron through the dung of the sow.

Treatment/prevention

Iron may be provided *ad lib* in the form of a liquid, lick, paste or powder on the floor. All these methods have disadvantages. Some piglets may not voluntarily consume the iron. The alternative is to dose the piglets with a fixed amount of iron. When given by mouth, the dose must be repeated several times. A better and possibly cheaper method, taking labour costs into account, is to inject the piglets with a suitable and safe iron preparation. It is *not* necessary to give iron at three days of age in spite of what is said by manufacturers.

In theory, the injection of iron will provide the piglet with a store of iron which is a massive overdose in relation to its needs at the time and which it will not be able to excrete from the body until such time as it starts to consume creep feed. Furthermore, 100 mg (1 cc) of iron per piglet is all that is necessary in most situations. The injection of 2 cc into the muscle of a small three-day-old piglet is painful and stressful.

Properly controlled trials carried out in several large herds in the North East of Scotland showed that 100 mg of iron dextran given subcutaneously at seven days of age was adequate for the prevention of anaemia. It must be pointed out that these trials were conducted in farrowing–rearing pens in which the sow's trough and the farrowing crate were made of ferrous materials. It is known that in licking the trough and the farrowing crate the sow ingests iron and this can increase the amount of iron in the sow's dung. Piglets may receive a certain amount of iron from this latter source. Delaying injection to seven days of age has an added advantage in the presence of neonatal diarrhoea. In fact, a piglet of four days of age or less which is suffering from any infectious disease should not be given iron by injection. This is because the defence mechanism of the piglet (reticulo-endothelial system) is at full stretch dealing with the infection, and saturation with iron particles will only heap insult upon injury. It is expecting too much of the body system to deal with two foreign invaders at one time.

Occasionally, a litter of pigs will become pale looking and anaemic in spite of iron being given by injection, perhaps more than once. The cause of this anaemia is unknown and affected litters usually recover rapidly when the pigs are weaned and forced to eat solid food.

FEEDING THE SOW

The nutrition of the sow is covered in detail in Chapter 11. The objective of the feeding policy in lactation is to cater adequately for milk production, to prevent excessive weight loss while suckling and to stimulate prompt appearance of heat and conception soon after weaning.

Sometimes, reduced feed level just after farrowing is advocated with a view to reducing risk of conditions leading to agalactia. While there may be a link between high feed level around farrowing and likelihood of conditions leading to agalactia under some circumstances, this relationship does not seem to hold in other situations. If too low feeding levels are practised around farrowing, these can cause increased restlessness in sows with good appetites just after farrowing and this can increase risk of overlying of piglets. It would seem sensible on those units where there is no evidence of a relationship between feed level around farrowing and incidence of conditions leading to agalactia to feed higher levels to sows with healthy appetites from farrowing.

If sows are tending to lose too much condition in lactation, then feed allowance should be increased. If the appetite of sows is too low to allow the sow to maintain reasonable body condition, attempts can be made to improve feed intake by any or all of the following methods:

- Feed more frequently.
- Feed *ad lib* during at least part of lactation.
- Feed wet rather than dry.
- Lower the temperature in the farrowing house (but not in the piglet creep).
- Feed a higher-energy diet.

CREEP FEEDING

It was established in Chapter 8 that sow milk yield reaches a peak around three weeks of age and thereafter declines to reach a low level by eight weeks of age. In the case of piglets to be weaned between four and eight weeks of age, it is essential that intake of supplementary feed be encouraged in order to achieve reasonable growth rate after three weeks of age and hence reasonable weight at weaning. This is shown in Figure 9·5.

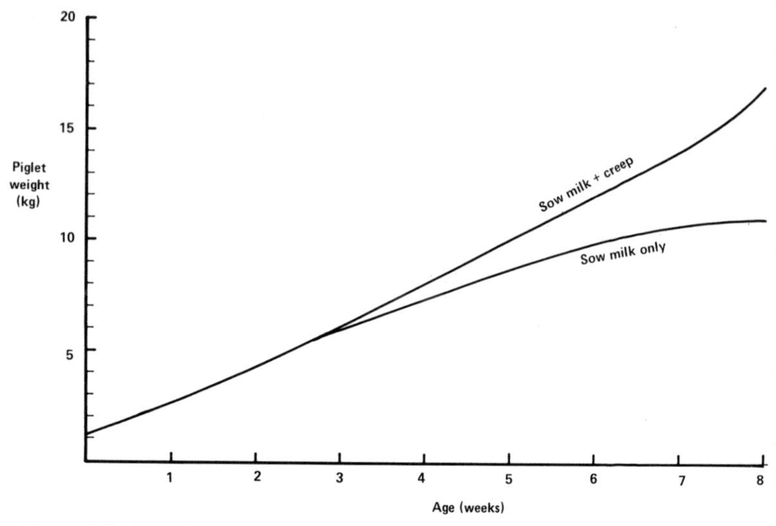

Figure 9·5. Growth of piglets on sow milk alone and on sow milk plus creep food.

Thus piglets require creep feed for two basic reasons. It is required to augment a milk supply which is inadequate for maximum growth, especially after three weeks of age. It is also required to accustom piglets to dry feed so as to encourage increased consumption by weaning and so reduce growth checks and digestive upsets at this stage.

TYPE OF CREEP FOOD

The type of creep food to be used depends on the age of piglet and, in particular, on the development of its digestive system. In the first two weeks or so of life the piglet can digest adequately only milk proteins (casein), milk sugar (lactose), glucose and fat. The enzymes necessary to digest starch, sugar (sucrose) and non-milk proteins develop only around two to three weeks of age. The development of these enzymes can be speeded up by encouraging the piglet to consume small quantities of these non-milk nutrients at an early age.

Thus, if increased creep-feed intake at an early stage is desired, the creep-feed should be very largely based on skim milk and added fat (preferably unsaturated fats such as maize oil or peanut oil) with small amounts of such non-milk nutrients as starch, sucrose and good quality non-milk proteins so as to encourage development of the mature digestive system. Such a diet should contain around 23 per cent crude protein, 1·2 per cent lysine and 13·5 MJ of digestible energy (DE) per kg. Around three to four weeks of age, this largely milk-based diet can gradually give way to a cheaper follow-on diet based on cereals (e.g. rolled oats, flaked or ground maize, ground barley), good quality protein (e.g. white fish meal, soyabean meal, microbial protein with about 5 to 10 per cent of dried skim milk retained), a small amount of sugar (sucrose), minerals and vitamins. This follow-on creep diet should have around 18 per cent crude protein, 0·9 per cent lysine and 13·0 MJ DE per kg.

It is usually very worthwhile to set out to maximise intake of creep food. A milk-based creep food will have a food-to-gain ratio (FGR) of 1 to 1 or better (1 kg of liveweight gain from 1 kg or less of creep food). The follow-on creep will have an FGR of 2 to 1 or better (1 kg of liveweight gain from 2 kg or less of creep food). It follows then that if the value of 1 kg of liveweight gain exceeds the cost of 1 kg of milk-based starter creep or that of 2 kg of follow-on creep, then it is worthwhile to set about maximising creep food intake. Usually, the value of piglet weight gain is worth almost twice as much as the food cost of producing it. Simple examples using current costs (May 1982) are shown in Table 9·9.

Table 9·9. Margin of piglet over creep food costs

Creep feed	*Cost per tonne*	*FGR**	*Cost per kg of liveweight gain (A)*	*Value of 1 kg of piglet liveweight (B)*	*Margin B minus A*
Milk-based starter (21–23% CP)	£480	1:1	£0·48	£1·05	£0·57
Follow-on creep (18–20% CP)	£200	2:1	£0·40	£1·05	£0·65

* FGR = kg creep per kg of liveweight gain.

It is particularly important to maximise intake of creep food if weaning early. Some may argue that too heavy a weaner will not grade so well, because of excessive fatness and that a piglet less well treated in terms of creep feeding during suckling will tend, after weaning, to catch up a piglet which has been well provided with creep food. While there is some truth in such arguments, they tend to be of less relevance in the practical situation than getting a well-grown pig by weaning as a result of good creep-feeding management. A well-grown pig at weaning which has consumed a large quantity of creep food will more readily adapt to the changes imposed by weaning and, the better the weight for age and the solid food intake at a given age, the greater the consideration which can be given to weaning earlier.

Careful attention to creep-feeding is likely to be of greatest benefit to those piglets which are receiving less milk from the sow.

Creep feed may be in the form of meal, crumbs or pellets, and it is important to make it as digestible, attractive and available to the piglet as possible. There is some indication that piglets initially find a meal with an open texture more acceptable than hard pellets. Good texture can be obtained by using open or flaky materials such as rolled oats or flaked maize. As well as good texture, inclusion of such products as dried skim milk, rolled oats and a small quantity of sugar (sucrose) help to increase acceptability. These desirable characteristics can also be improved by avoiding dustiness and ensuring that creep food is always provided in a fresh state.

Few stockmen provide fresh creep every day which is the ideal and will result in much higher consumption at a stage when feed efficiency is very high. All too often, the creep is put into a hopper and left there to go stale, be fouled, and attract flies before being looked at again and fresh material added. Creep-feeding management certainly leaves much to be desired and more attention would be likely to pay handsome dividends. 'Little and often' must be the rule with fresh creep given each day. Not only does this practice ensure that creep food is always fresh, but the daily arrival of new material in the creep feed trough serves to stimulate the inherent curiosity of the piglet in the new material and this helps to encourage intake. On adopting the 'little and often' system of feeding stockmen will be pleasantly surprised at the increases in consumption and in weaning weights which can be achieved. Another useful practice which helps to increase creep-feed consumption at an earlier stage is to scatter a small amount of dried milk powder over the creep-feed pellets or crumbs. This attracts the piglets and helps to stimulate earlier consumption of solid food. This is especially useful with early weaning as it helps to reduce growth checks and digestive scour after weaning.

There is also a case for supplementary feeding at a very early stage to help out litters which are obviously going short of milk. In the case of a sow suffering from post-farrowing agalactia, feeding limited quantities of liquid milk substitute four to six times a day for a few days can often help keep a litter thriving until the sow

Plate No. **34**

Shallow circular troughs make creep food more obvious and available to piglets and can accommodate a fairly large litter feeding together.

fully recovers and can cope adequately herself. If some sows continue to milk poorly, supplementary liquid feeding can give way to inclusion of a dry milk-based pellet in the creep trough from about five to seven days after farrowing. Again, this dry feed should be given fresh each day.

Since piglets prefer to feed together, the feed trough provided should be large enough to allow all piglets within the litter to feed at the same time. A fairly heavy, shallow circular trough is suitable at the start. Being shallow, the creep food is more obvious and available to piglets and a circular feeder can accommodate a fairly large litter at the outset (see Plate 34). Once creep feeding has been established, a larger feeder can be provided but the 'little and often' principle must be maintained with fresh creep being added daily according to appetite.

If the creep feed is constantly being fouled by the piglets, this indicates that it is wrongly situated. Moving it nearer the piglet lying area will usually solve this problem. Piglets should always have available a plentiful supply of fresh clean water and it is more convenient and hygienic to provide this using a nipple drinker.

Chapter 10

WEANING, MATING AND PREGNANCY MANAGEMENT

THE PRODUCTIVE periods in the life of the sow are pregnancy and lactation. The sow is unproductive from selection (usually at slaughter weight of contemporaries) to first conception and between each weaning and subsequent conception. During these periods, the sow is neither developing piglets in the uterus nor nursing them and is consuming food and using other resources for no return.

PRODUCTIVE AND UNPRODUCTIVE PERIODS

Thus, we can identify quite distinct productive and unproductive periods in the life of the sow. These are:

1. Pregnancy 2. Lactation	} Productive periods
3. Selection to first conception 4. Weaning to conception 5. Final weaning to culling or death	} Unproductive periods

The duration of pregnancy can vary between sows but the herd average is virtually constant. The length of lactation can be varied by varying the age at weaning. Age at weaning and the ratio of productive to unproductive periods dictate the number of litters possible per sow per year which is a crucial component in the profitability of weaner production (see Chapter 1). Thus, in order to increase number of litters per sow per year, attention must be focused on both age at weaning and the length of the unproductive period. It is the length of the unproductive period and methods of reducing this which form the basis of this chapter.

NUMBER OF LITTERS PER SOW PER YEAR— CALCULATION

Number of litters per sow per year is rightly regarded as a factor of crucial economic importance. In any calculation of the efficiency of weaner production, whether on the basis of one herd, a sample of herds or nationally, this component is always featured. However, the basis for this calculation is seldom defined. If one ignores one or other, or both of the unproductive periods 3 and 5 (that is, selection to first conception and final weaning to culling or death), one can, at a stroke, produce much more attractive figures in terms of litters per sow per year.

So the figure for litters per sow per year means very little unless one knows the basis of calculation, and especially whether or not unproductive periods 3 and 5 above have been included in the calculation.

Probably the fairest basis of calculating litters per sow per year is to count the gilt as a member of the sow herd from a standard weight, say, 90 kg liveweight, and treat all sows as members of the herd up to culling or death.

UNPRODUCTIVE PERIOD—THE EXTENT

To get some measure of the total length of the unproductive period in practice, one can refer to a variety of survey figures. Those of the Cambridge University Pig Management Scheme are useful as these examine levels of sow productivity in relation to different ages at weaning. The age at weaning categories are less than twenty, twenty to twenty nine, thirty to thirty nine and forty days and over. The mean age at weaning within these categories are 17, 23, 35 and 44 days respectively and the length of the unproductive period is as shown in Table 10·1.

TABLE 10·1. Calculation of length of unproductive period per year

	Less than 20	20–29	30–39	40 and over
Age at weaning (days): range	Less than 20	20–29	30–39	40 and over
mean (A)	17	23	35	44
Number litters per sow per year (B)*	2·31	2·24	2·05	1·94
Total days per year: pregnant (B × 114)	263	255	238	221
lactating (B × A)	39	52	72	85
productive (C)	302	307	310	306
Total unproductive days (D) i.e. (365 minus C)	63	58	55	59
Allowance for normal weaning to first heat period of 7 days (E) i.e. (7 × B)	16	16	14	14
Net unproductive days (D minus E)	47	42	41	45

*Calculated by including gilts from date of first service

Source Ridgeon, R.F., Cambridge University Pig Management Scheme Results, 1981.

It can be seen from Table 10·1 that after an allowance of seven days is made to cover weaning to first heat period, sows weaned at different ages spend a considerable proportion of their time in an unproductive state, that is 47, 42, 41 and 45 days per year for pigs weaned at a mean age of 17, 23, 35 and 45 days respectively. The associated production loss is shown in Table 10·2.
10·2.

TABLE 10·2. Loss of weaners caused by the unproductive period

	Age at weaning (days)			
	17	*23*	*35*	*44*
Length of normal reproductive cycle (days):				
pregnancy	114	114	114	114
lactation	17	23	35	44
weaning to conception	7	7	7	7
Totals (A)	138	144	156	165
Theoretical number of litters per sow per year $\left(\frac{365}{A}\right)$	2·64	2·53	2·34	2·21
Actual number of litters per sow per year (from Table 10·1)	2·31	2·24	2·05	1·94
Lost output (litters per sow per year)	0·33	0·29	0·29	0·27
Loss of potential weaners (at 9 per litter)	2·97	2·61	2·61	2·43

So that, on the basis of the above figures, from the time of first service as gilts, sows are unproductive for between 10 and 13 per cent of the time and this results in a loss of between two and three weaners per sow per year.

Not only are potential piglets being lost because of this long unproductive period but expensive resources such as feed, labour, power and accommodation are being used for no return. This is therefore a problem area on most pig units which demands urgent attention.

UNPRODUCTIVE PERIOD—THE REASONS

The possible reasons for these long unproductive periods each year are several:

- Failure of sows to show oestrus activity promptly or at all after weaning (anoestrus).
- Failure to detect oestrus or heat.
- Failure to conceive.

- Embryonic losses and abortion.
- Delay between weaning and culling.

While sows should show oestrus activity within seven days of weaning, findings of a survey based on a large number of herds in France as shown in Figure 10·1 tell a different story.

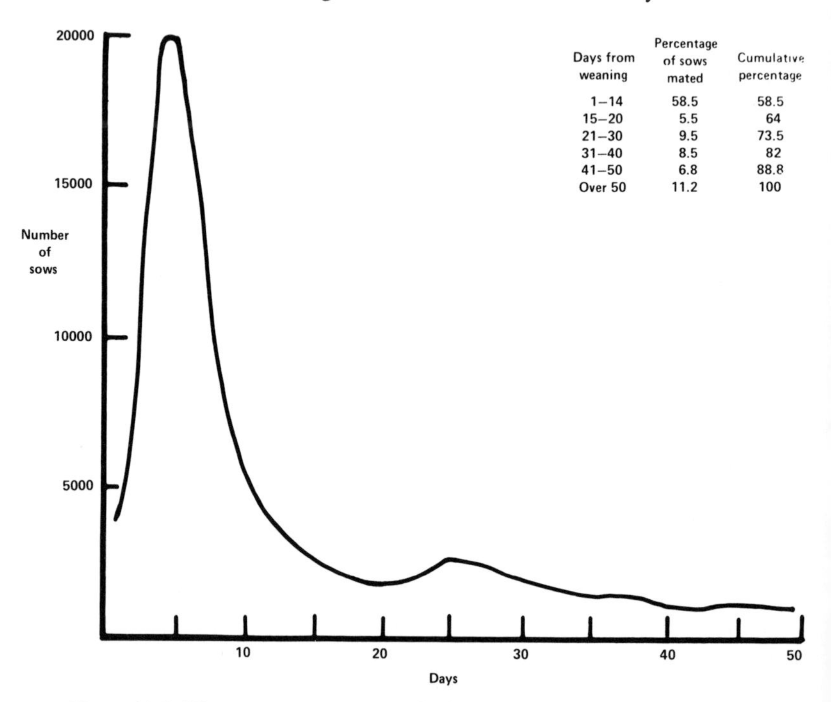

Days from weaning	Percentage of sows mated	Cumulative percentage
1–14	58.5	58.5
15–20	5.5	64
21–30	9.5	73.5
31–40	8.5	82
41–50	6.8	88.8
Over 50	11.2	100

Figure 10·1. Weaning to mating interval (days).

Source: Legault, C., Dagorn, J., & Tastu, D. (1975). French Survey Results.

Thus, less than 60 per cent of sows were mated within fourteen days of weaning and a considerable time elapsed before a proportion of sows were mated after weaning. It is likely that failure to come in oestrus (heat) promptly after weaning and failure to detect heat are both implicated in this problem. Conception rate to first service after weaning varies between herds from 60 to almost 100 per cent, while total loss of embryos and abortions can also be problems.

Very many factors are involved in these inefficiencies which can lead to long unproductive periods in sows. These include genetic factors, disease, feeding, housing, behavioural, stockmanship and management considerations.

Before examining the role of these factors in contributing to reproductive problems and methods of minimising the breeding shortfalls of the sow, the process of normal reproduction following weaning will be outlined briefly.

REPRODUCTIVE PROCESSES

Hormones

During lactation, certain hormones are involved in milk production (prolactin) and in milk 'let-down' (oxytocin). While suckling a litter, the sow is normally anoestrous, i.e., she shows no sign of breeding activity, because as long as she is lactating, the hormones which stimulate breeding activity are either not produced or are not released from their production sites. Thus, lactation is said to deter breeding activity (lactational anoestrus). When the sow is weaned there is almost an immediate response in terms of breeding activity.

Mating in lactation

Some exceptions to anoestrus in lactation do arise. For instance, an apparent heat can sometimes be observed about two days after farrowing (post-partum heat) but no ova are shed at this heat. Therefore, service at this heat will not result in pregnancy. It just so happens at this time that the hormone balance can be very similar to that associated with a true heat and therefore the sows can show all the signs of being in heat and will allow service.

Some attempts have been made in recent years to break the lactational anoestrus by husbandry modifications such as grouping sows and litters about three weeks after farrowing, boosting feed intakes and running a boar with such groups. Such practices have resulted in a proportion of sows coming in heat, being served, conceiving and subsequently producing perfectly normal litters. However, the success rate in terms of proportion of sows conceiving while still lactating varies markedly from farm to farm and even on the same farm. Since the precise reasons for such variations in success rate are not yet known, the technique of 'sow service in lactation' cannot yet be recommended for practical application.

Breeding activity after weaning

With the influence of lactation and suckling removed at weaning, in the normal healthy sow under the right conditions, there is a rapid growth of the follicles in the ovaries which contain the ova or

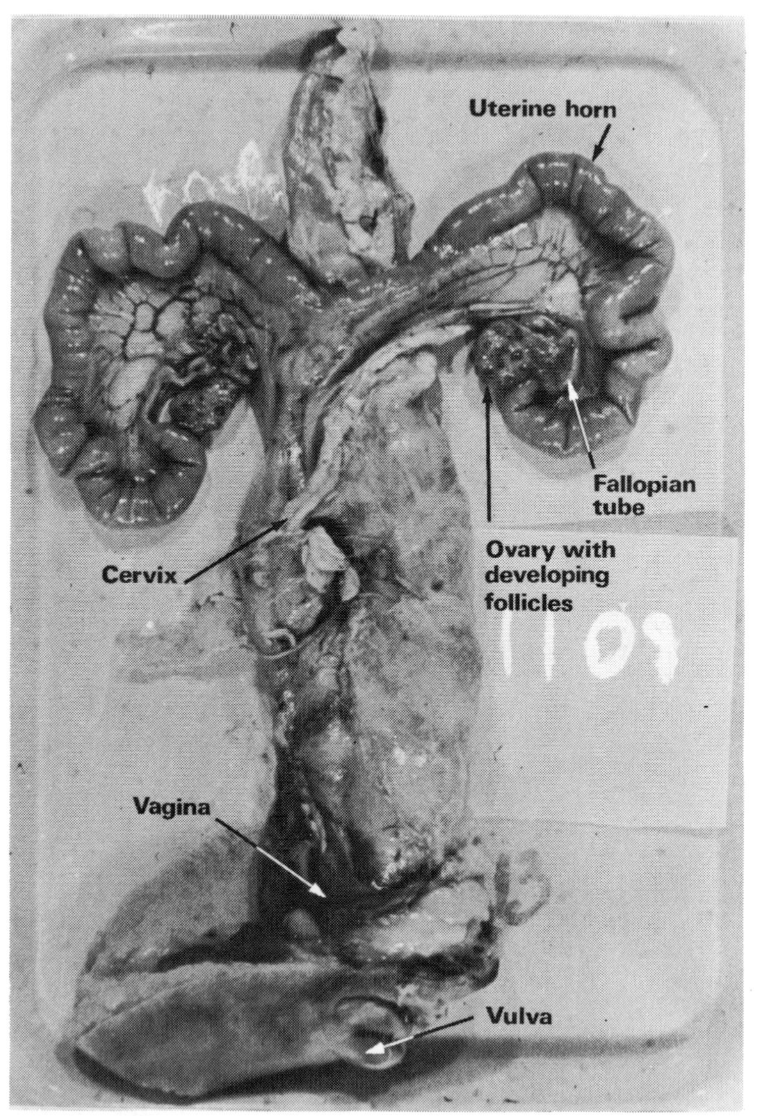

Plate No. **35**

Reproductive tract of a gilt showing ovaries and developing ova, fallopian tubes, uterus, cervix and vagina.

eggs. These grow to a stage where, normally, four to seven days after weaning, the eggs are shed from the follicles into the fallopian tube (see Plate 35). The shedding of the eggs in this way is termed ovulation. Just prior to ovulation, the sow comes into oestrus or heat.

OESTRUS AND ITS DETECTION

Oestrus is the period during which the sow will allow service by a boar and can last from one to three days. As oestrus approaches, the vulva swells and reddens and these signs are most noticeable in gilts. Other signs of heat include mounting of other sows or allowing themselves to be mounted, pricking of ears in prick-eared types of sow such as the Large White. Sows in heat are also more restless (they may be the only sows standing up at times when the rest are asleep) and will often issue characteristic barks associated with heat. They will also show the 'standing' reflex, i.e., they will stand rigidly when pressure is applied to the back. At the start and towards the end of oestrus, the sow will show the 'standing' reflex only in the presence of a boar. For a period at the height of oestrus, however, many sows will show the standing reflex when pressure is applied on the back when the boar is absent.

The 'standing' reflex is the most foolproof sign of heat. Sows more readily show the 'standing' reaction to back pressure when a boar is in the vicinity, less than 50 per cent of gilts in heat actually showing the standing reaction in the absence of the boar (see Table 10·3). Results of French work indicate that the smell of the boar is the major factor in inducing sows to show the standing reflex although sight, sound and contact are required to increase the incidence of standing reactions in oestrous sows up to 100 per cent. This is illustrated in Table 10·3.

TABLE 10·3. Percentage of gilts in heat showing the standing reflex in response to various boar stimuli

Boar stimuli	*Percentage showing standing reflex*
None	48
Smell and sound	90
Smell, sound and sight	97
Smell, sound, sight and contact	100

Source: Signoret, J. P. and Du Mesnil Du Buisson, F. (1961).

The 'boar smell' factors which induce oestrous sows to show the standing reaction are the pheromones produced in the submaxillary or preputial glands. The most likely pheromone exerting this influence is androstenol which is concentrated in the submaxillary gland. The boar, in champing its jaws in the presence of the sow, probably releases this pheromone. The chemical androstenol has now been synthesised and is available commercially in aerosol form. A spray of this material is useful when no boar is available to increase detection rate in sows. While this technique is less effective than the boar in helping to detect sows in heat because it obviously lacks the sight, contact and sound effects of the boar, it is useful in those herds entirely dependent on AI where no service boar is kept.

Another important point in heat detection is that it is not so much the boar which detects the oestrous sow but that the sow on heat will attract the boar's attentions and stimulate his interest. This interest can then be detected by the stockman. It is easier for the sow to attract the boar in this way if he is taken up the front of sows in dry stalls rather than behind them.

Of course, it is preferable for sows suspected of being in heat or due to be in heat to be taken to the boar pen.

Sows differ in the ease of heat detection. While the great majority are fairly easy to detect as being in heat, a few impose particular difficulties. Some sows, for instance, will show preferences for particular boars. Such sows may fail to show any of the behavioural signs of heat in the presence of one boar but demonstrate all the obvious signs when in contact with another boar. Similarly a shy, timid sow may show no signs of heat when taken to the boar's pen but when she is taken back to her own pen, the signs of heat will become obvious.

Thus, while perhaps about 90 per cent of heats can be detected fairly readily, the remaining 10 per cent may be more difficult to detect. Thus, a dedicated and observant stockperson with adequate time available is required if high oestrus detection rates are to be achieved.

SYNCHRONISATION OF OESTRUS

Synchronisation of oestrus in a group of sows has several objectives related to easing management and increasing productivity. A batch of sows in oestrus at the same time facilitates the use of AI and reduces its cost. A futher objective is to synchronise farrowing, although the variation between sows in gestation length can reduce the overall effectiveness of this approach.

As outlined in Chapter 5, the progestagen allyl trenbolone (Regumate, Hoechst UK Ltd) gives promise of being useful for synchronising oestrus in gilts which have already had one 'standing' heat. The product should be added to the daily ration at a rate of 20 mg daily over an eighteen-day period. About eight days after the last dose of Regumate, gilts in a treatment group will come into heat within two to three days of each other. The most used natural methods for sychronising oestrus in practice are batch weaning in sows and boar influence in gilts. The latter technique involves introduction of a boar to a group of maiden gilts at the appropriate stage (around 170 days of age for crossbred gilts) and this technique has been discussed fully in Chapter 5. Batch weaning is a widely adopted means of synchronising oestrus although its effectiveness is often reduced because of variation in the weaning-to-oestrus period.

TIMING OF MATING

Fertilisation takes place at the ampullary-isthmic junction in the fallopian tube (See plate 35). The objective is to arrange mating so that sperms and eggs arrive at that site at the same time. This ensures fresh sperm and fresh eggs at fertilisation and this makes for optimum fertilisation rate and subsequent development.

The eggs, after being shed from the ovary, retain their vitality for a very short period and, if service is too late, then abnormalities in fertilisation and abnormal embryos can arise. The sperm remain viable for a longer period than the egg but service must not take place too soon before the eggs are shed. After service, two to four hours must elapse to allow sperm to be exposed to fluids from the uterus and oviduct as the effects of these fluids on the sperm facilitate fertilisation. This process is called capacitation of sperm. The importance of proper timing of service within the heat period is illustrated in Figure 10·2.

The different stages in oestrus are indicated in Figure 10·2 and it can be noted that ovulation occurs towards the end of heat. The sow can be served at any time over a fairly protracted period but there is an optimum time for service at which both conception rate and litter size are maximised. This timing ensures that sperm and eggs arrive fresh at the same time at the site of fertilisation. If service takes place too soon in the heat period, the sperms may be too old for optimum results by the time the eggs are shed. On the other hand, if service takes place too late, then the eggs will have been adversely affected by ageing before the sperms arrive.

Thus, optimum service timing will result in maximising both litter size and conception rate and failure to achieve this optimum timing can often result in quite marked depressions in both litter size and conception rate. Because length of the 'standing' period of heat and timing of ovulation varies, unfortunately it is not possible to predict the timing of this optimum point so that service can be arranged to coincide with it. The practical trick used for overcoming this problem, of course, is to double or multiple serve within this period so as to increase the chances that one of these services will occur at the optimum time.

The advantages of double over single service are shown in Table 10·4.

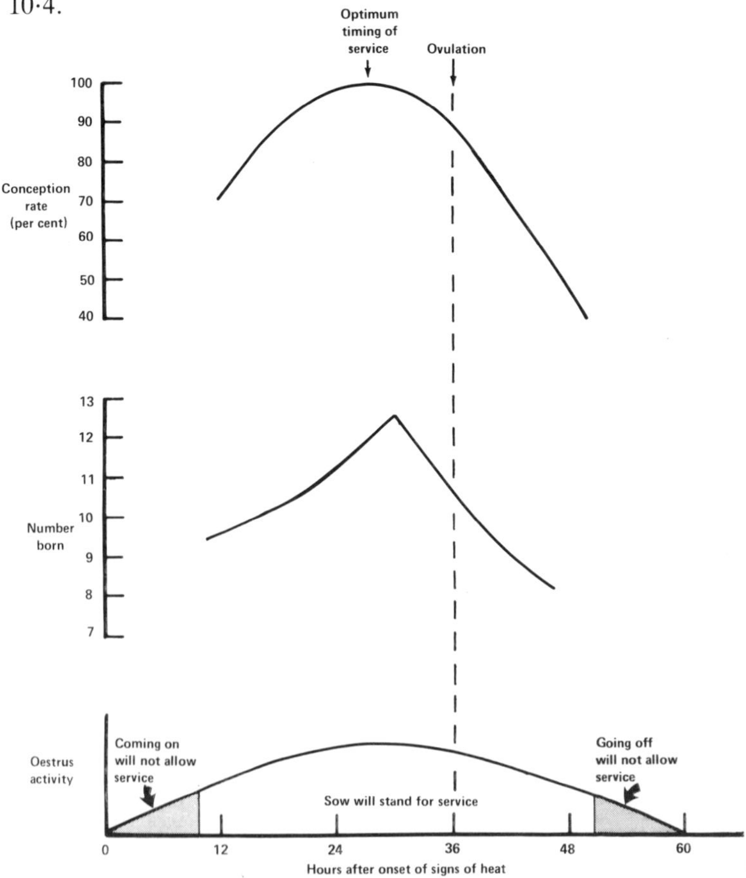

Figure 10·2. Stages in oestrus or heat and optimum timing of service.

TABLE 10·4. Effect of number of services

	Number of services	*1*	*2*
Return rate	Farm 1	29	15
(per cent)	2	39	16
	3	20	10
Litter size	Farm 1	11·5	11·6
	2	7·9	8·9
	3	9·3	10·7

Thus, proportion of sows failing to conceive to first service was doubled as a result of single relative to double service while litter size was also increased by double service.

In theory, further dividends are possible by adopting triple service, such as early morning, late afternoon and early the following morning. Obviously, arranging triple or quadruple service is more time-consuming and more demanding on boars. But, for producers with a low litter size problem, it is one way of trying to improve the situation and getting a useful boost in conception rate at the same time.

It is well recognised that if there was some basis for determining optimum timing of service, then there would be no need for multiple service and a single service done at the optimum time would suffice. There is some evidence that the optimum timing of service can be determined by monitoring changes in the conductivity of the vaginal mucus and this technique has been automated (see Plate 36).

Such equipment has been in use over a fairly lengthy period on a large commercial unit in Aberdeenshire. The manager has reported the following results.

	Previous practice (Double service on Consecutive days) (Jan–April 1976)	*Present practice* (Single service. Timing dictated by Walsmeta) (Jan–April 1977)
Number sows*	183	174
Conception rate	88·5	93·1
Litter size	10·9	11·6

*No gilts, but sows only, involved in this exercise.

Source: Bridgeford, P. (1977). Arnage Pig Unit, Aberdeenshire.

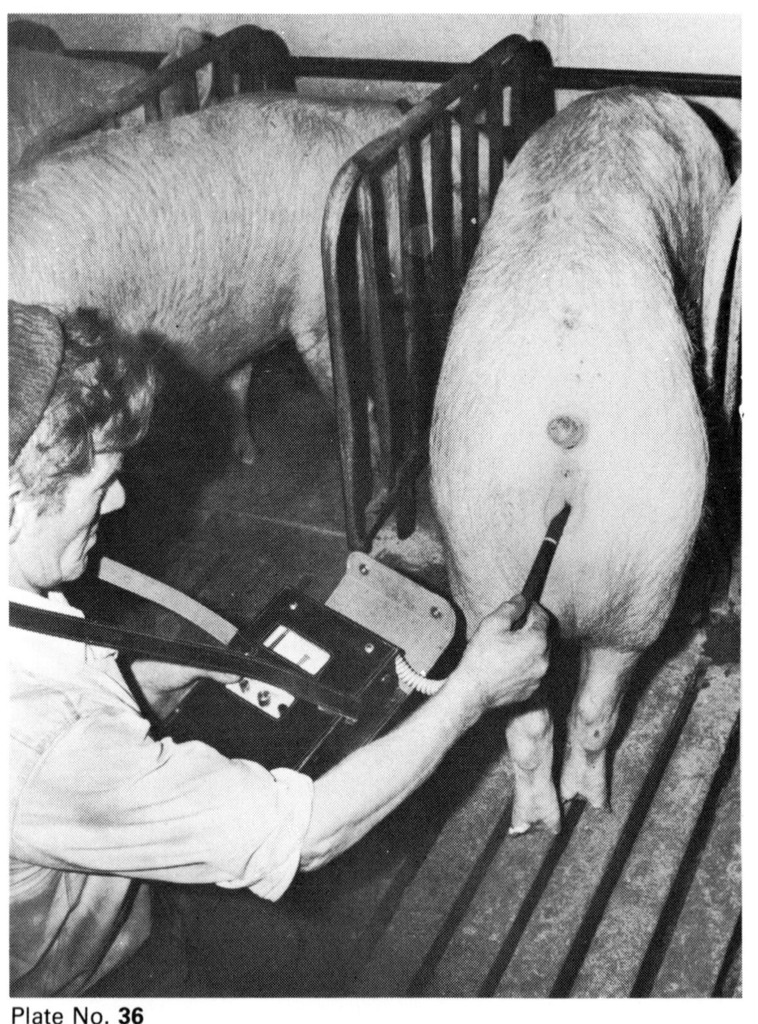

Plate No. **36**

Checking sows for oestrus or heat using the 'Walsmeta' device.

Results from the practice of single service as timed by the 'Walsmeta' device are encouraging and this is now established as the standard practice on this farm. Much of the success of the equipment on this farm can be attributed to the skill and sensitivity of the operator. This fact needs emphasis as such devices are certainly no substitute for good stockmanship but rather a potentially useful management aid for skilled operators.

Obviously, considerable economies in number of boars to be kept could be achieved by adopting a single service policy at the optimum time rather than a double or multiple service policy.

ARTIFICIAL INSEMINATION

Following appropriate dilution, one ejaculate from a boar can be used to inseminate up to twenty sows, the exact number of semen doses obtained depending on the quality of the semen. Thus, semen from one boar can be used to inseminate up to 3,000 to 4,000 sows per year and this helps to exploit the use of high-quality sires.

The rules for timing of insemination to obtain optimum results are identical to those for natural service as outlined in the previous section. The advantages of double over single insemination are similar to those for double over single service.

Insemination is carried out by using a catheter moulded in the shape of a boar's penis. If the sow is at the correct stage for insemination, the catheter will become locked in the cervix, this being a useful check that the sow is at the proper stage for insemination. Conception rate and number born following insemination should be as good as with natural service provided timing is correct, semen is of good quality and is allowed to run into the uterus by gravity rather than being forced in. In practice AI tends to give poorer results than natural service in terms of both litter size and conception rate mainly because of poorer timing of service and possibly also because of insufficient care taken with technique and hygiene.

Up to recently, boar semen could only be used fresh but the problems of deep freezing and storage have now been at least partly overcome.

AI is very useful for the farmer with too few sows to justify the keeping of a boar and it is also used to an increasing extent by farmers operating a closed herd policy to introduce new genes without running the risk of introducing disease. Since most boars standing at AI stations are of a high calibre from the point of view of efficiency of production, use of such boars through AI is also useful in helping to bring about herd improvement.

FERTILISATION, IMPLANTATION AND GESTATION

Following ovulation, the follicle from which the egg was shed is occupied by the yellow body or corpus luteum. This produces the hormone progesterone which helps to maintain pregnancy if fertilisation has taken place. If fertilisation has not taken place, then the corpus luteum will regress after about fourteen days, a new batch of ova will develop and these will be shed, on average, twenty-one days after the previous ovulation. Such failures to conceive add considerably to the length of the unproductive period when calculating the sow's annual output. Thus, maximising conception rate to first service is very vital for increasing sow output.

While the sow, on average, sheds between seventeen and twenty ova at each ovulation, only about eleven of these will end up as viable piglets at birth. The shortfall between ovulation and birth may arise because of failure of fertilisation, failure of implantation of the fertilised ova in the wall of the uterus and because of embryonic deaths.

Embryonic losses can be as high as 40 per cent and most of these losses occur in the first thirty days of gestation. This period coincides with implantation of the ova in the wall of the uterus, which takes place about fourteen days after mating, and the early critical stages in development of the embryos. While some of the early embryonic losses may be caused by genetic defects, the young embryos tend to be very fragile and sensitive to any adverse effect in this critical time. Consequently, the sow must be well cared for if such losses are to be minimised.

While most of the embryonic deaths will occur in this early period, some deaths may occur around day 70 of gestation because, at this stage, the placenta stops growing whereas the growth of foetuses is just beginning to accelerate. Thus, an increasing mass of pigs is competing for a static area of placenta and the smaller foetuses may suffer to such an extent that they die.

HOW TO AVOID EMBRYONIC LOSSES

Among the steps which should be taken to minimise embryonic losses are the following:

- Take all possible steps to minimise risk of introducing disease.
- Ensure proper timing and frequency of service (avoid fertilisation of an ageing egg).
- Avoid excessive food intake after service as this is associated with losses of embryos.
- Avoid excessive environmental temperature in early pregnancy as this can increase losses of embryos.
- Avoid imposing undue stress on sows in early pregnancy, e.g.

by mixing them with others. The more peaceful and contented they are in early pregnancy, the lower the embryonic losses. This was indicated by Polish work in which diets fairly high in roughage were given in early pregnancy and this resulted in improved numbers born. This was attributed to the fact that sows were kept more contented, resulting in improvements in fertilisation and implantation rates. The feeding of straw to sows in dry sow stalls could produce a similar effect. Certainly sows eat considerable amounts when given the opportunity.

- Adopt a crossbreeding policy since crossbred sows suffer fewer embryonic losses than purebreds.
- Avoid very early weaning since it has been demonstrated that embryonic losses increase with earlier weaning. This is thought to be due to the fact that the earlier piglets are weaned the less well prepared the uterus will be to accept and develop a new litter.

FACTORS AFFECTING REPRODUCTIVE PERFORMANCE

In covering the normal processes of reproduction some mention was made of factors which affect these normal processes. In this section, the effects of more specific factors on reproduction are examined. These factors include:

1. Feeding.
2. Housing.
3. Season.
4. Boar.
5. Age at weaning.
6. Parity.
7. Genotype.
8. Genital and locomotor disorders.
9. Disease.

1. FEEDING

The objective of feeding to achieve a high level of breeding efficiency is to stimulate (i) prompt breeding activity after weaning, (ii) a well-exhibited oestrus so that it can be readily detected, (iii) an adequate ovulation and fertilisation rate and (iv) good survival of embryos.

An important factor which has emerged very clearly from the considerable amount of research work carried out on sow nutrition is that one cannot consider the effects of feeding in one period as being independent of the effects in another. Thus, while reproductive performance can be influenced by the effects of feeding from weaning onwards, performance can also be affected by the carry-over effects of feeding in the previous pregnancy and lactation.

These nutritional factors can influence the production and release of reproductive hormones and the effect of these on the reproductive organs, such as the ovary and uterus. The effects of nutrition on reproductive performance of the gilt have been covered in Chapter 5, while the role of feeding in affecting breeding performance in general is dealt with in Chapter 11. Here it will suffice merely to summarise the main effects of feeding on breeding.

HOW FEEDING AFFECTS BREEDING

From the physiological point of view, there is no need to reduce either feed or water at weaning in order to dry off the sow. The most effective way to dry off the sow is to allow milk to accumulate in the udder and the resulting pressure on the milk-producing cells will very quickly shut off milk production. There is, therefore, no justification for starving sows for 24 hours after weaning except possibly for sows in overfat condition where a sudden stress, such as starvation, can help to trigger oestrus.

Adequate intakes of protein are required in lactation in order to ensure prompt oestrus and ovulation after weaning (see Chapter 11, Table 11·2). Low protein intakes in lactation can cause excessive delays in heat and conception after weaning, especially after the first lactation. This problem may be associated with excessive weight loss in lactation and thin sows at weaning.

Continuation of high food intakes after weaning can result in earlier and more concentrated appearance of heat and conception after weaning, especially after the first lactation and more particularly in gilts which are much reduced in condition by weaning (see Chapter 11, Table 11·3). Effects of continuation of high feed levels after weaning are less noticeable in older sows but it is likely that a proportion of thinner sows in particular, will benefit from such a practice. It must be kept in mind that all sows and gilts continue to lose weight for a few days after weaning, so poor condition at weaning is further aggravated by this continued weight loss afterwards.

It is important that when high feeding levels are practised following weaning these should be discontinued almost immediately after the last mating as high feed levels after mating increase embryo losses, especially in gilts.

Flushing of sows, i.e., increasing the feeding level before or on the first day of mating, is unlikely to increase ovulation rate and numbers born in sows. When gilts which have been reared on a restricted plane of feeding have their food intake boosted by 50 to 100 per cent some ten days before mating, this can result in increased ovulation rate and an increase in numbers born in some herds.

There is some evidence that extra vitamins at weaning can prove useful. Some producers who include liberal quantities of white fish meal in the diet for a few days after weaning claim substantial benefits in terms of short weaning-to-conception intervals and good subsequent litter size. For the same reason, there is much support on many units for the practice of administering an injection of multivitamin (both fat and water soluble) to all sows at weaning. Recent work carried out at Seal Hayne Agricultural College on providing more biotin to sows at weaning has produced useful reductions in weaning-to-conception interval in the College herd.

Keeping sows more contented in early pregnancy by providing diets with a higher than normal roughage content may be helpful for reducing losses of embryos. This may be achieved simply by offering straw to dry sows.

While the adverse effects of poor condition at weaning can be reduced by offering liberal amounts of food from weaning, especially after the first lactation, it is better to isolate the causes of such poor condition in the first place and take appropriate avoiding action. One can first offer more food in lactation, if necessary by more frequent feeding or else by *ad lib* feeding during at least part of lactation. If the problem is low appetite in lactation, then one should examine feed intake in pregnancy, and reduce this if excessive, for the higher the intake in pregnancy, the lower the appetite in lactation. Low appetite in lactation is often a particular problem in gilts and the cause can often be excess feeding in pregnancy resulting in overfat gilts at farrowing. It must be kept in mind that the gilt, being a smaller animal, has a much lower maintenance requirement than the older sow (see Chapter 4, Figure 4·2). So considerably lower feed levels should be offered to gilts than to older, heavier sows in pregnancy.

Other steps one can take to increase food intake in lactation and so avoid severe weight loss and poor condition at weaning include:

- Lower the temperature for the sow in the farrowing house by concentrating heat in the piglet creep or micro-environment

and/or by lowering farrowing house temperature after the first few critical days of life.
- Feed a higher energy diet.
- Feed wet rather than dry.
- Feed pellets rather than meal.

2. HOUSING

Various factors pertaining to housing, the climatic environment and penning arrangements affect breeding activity and herd-breeding performance. These include the following:

- Single versus group housing of sows at weaning.
- Degree of contact between boars and sows.
- Ease of moving sows and boars to facilitate heat detection and service.
- House temperature.
- Adequacy of service area.

Single versus group housing of sows

There is much controversy over whether sows are best housed singly or in groups at weaning for maximising reproductive efficiency. This controversy is not surprising for it is likely that the relative advantages and disadvantages of housing singly and in groups will vary with the particular circumstances prevailing. On being mixed at weaning, sows will fight in order to develop the peck or social order. This imposes stress on the sows and, while a certain amount of stress on sows is desirable at this time since it is likely to provide a stimulus to breeding activity, this can be overdone and result in bullying, injury and semi-starvation in some sows. The level of stress imposed as a result of fighting will depend on the number in the group, the space provided, the distractions available and the competition for feed. Thus, if at weaning, a large number of sows are grouped in a confined space in a concrete-floored pen without bedding and group fed on low feed levels in a limited feeding area, then this will make for intense competition between sows, and some or all are likely to be overstressed and performance will suffer.

If, on the other hand, sows are divided into fairly small groups at weaning, placed in a large pen with straw bedding and fed by scattering liberal quantities of sow nuts on the floor (or by providing individual feeding facilities), then this is an entirely different situation. Fighting will certainly take place in order to

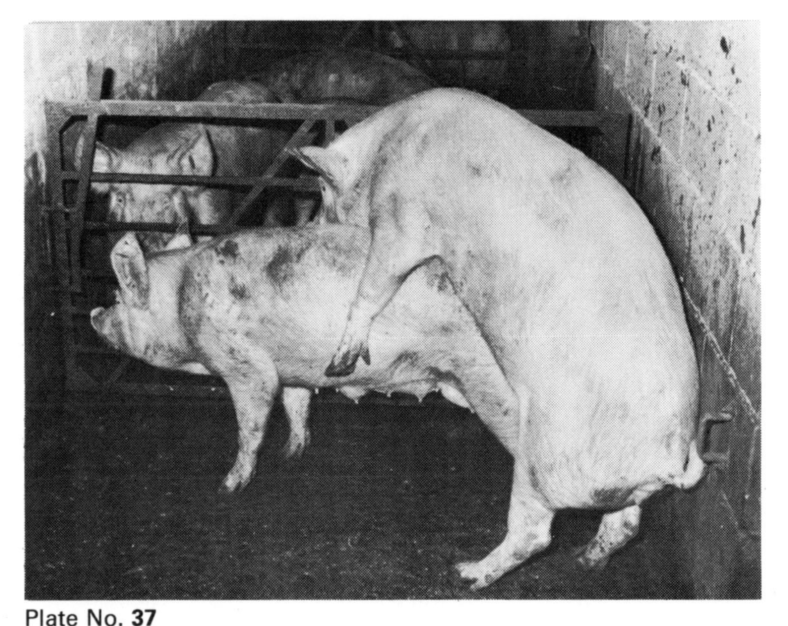

Plate No. **37**

Mounting behaviour in recently weaned sows. Note the boar in the adjacent pen. The signs of heat are more noticeable when newly weaned sows are group-housed.

develop the peck order but it will be less obvious and will have a less adverse effect on individual sows since those timid ones will be better able to keep out of the way of their superiors and at the same time obtain adequate food, water and comfort within the pen. The lower level of stress imposed in this situation is likely to have a stimulating effect on breeding activity following weaning.

It is noticeable, therefore, that one can have wide extremes in terms of the degree of stress imposed on sows at weaning as a result of grouping and these can produce very different effects in relation to breeding activity.

Heat detection is easier among group-housed sows than among those housed singly since they are able to interact with each other and these interactions, including mounting behaviour, will make it easier for the stockman to detect sows in heat. It is also easier in the group situation to arrange for more intimate daily contact with boars either by taking the boar to the sow pen or vice versa in order to increase efficiency of heat detection.

Another possible advantage of group housing is the fact that sows are able to huddle together to keep warm and therefore will have a lower feed requirement or can stand a lower temperature level than sows housed singly. Because they have a lower feed requirement for maintenance, they are likely to recover body condition more quickly after weaning than will sows housed singly on similar feed levels and subjected to the same environmental temperature.

However, in situations where one cannot prevent bullying and provide comfort and reasonably fair shares of food and water for all at weaning, individual penning is preferable.

If sows are group housed at weaning, it is desirable to keep them in such groups until after the three-week return date. By this time, implantation of the embryos should have taken place. On the other hand, if the sow has failed to conceive to first service, then 'returns' are usually easier to detect in a group-housing situation. About four weeks after service, and hopefully, conception, the sow is best housed individually to remove her from competition, in order to facilitate pregnancy diagnosis and to allow rationing according to condition.

Contact between boars and sows

It has been outlined earlier that the boar has a stimulating effect on breeding activity in the sow. This is mainly through the chemical messengers or pheromones produced by the preputial or sub-maxillary glands. Thus, it is the smell of the boar which has the main stimulating effect but the sound, sight and contact with the boar all help to maximise this effect. The presence of the boar also helps in detection of oestrus as less than half of sows in oestrus may show the typical signs of heat in the absence of the boar (see Table 10·3).

Thus, it is important in housing design to arrange for newly-weaned sows to be subjected to maximum boar influence. Boars should be housed in pens adjacent to newly-weaned sows, and some producers have now arranged air inlets above boar pens so that maximum boar smell is distributed over pens housing newly-weaned sows. It must be kept in mind that the boar pheromones are volatile and will circulate more effectively in warm than in cold housing. However, too stuffy an atmosphere with a high concentration of ammonia may tend to mask the boar effect.

If sows are slow to come in heat, the boar in closest contact with them should be changed as sows can become accustomed or 'habituated' to a given boar and receive a fresh stimulus when given contact with another boar. It is often worthwhile placing

Plate No. **38**

Newly weaned sows should be housed adjacent to a boar to provide the necessary stimulus to come in heat promptly.

such trouble-some sows along with a strange boar and this can often produce the desired effect and stimulate them to come in heat. In theory, sows which fail to come in heat after weaning (true anoestrous) should respond well to hormone therapy. In practice, however, the response may vary from very good to poor, even in the same herd. Nevertheless , PG600 (Intervet), containing follicle stimulating hormone (FSH) and luteinising hormone (LH), when given within a few days of weaning, has been found useful in some cases, especially in the case of first-litter gilts.

Ease of movement of sows to boar

It is very important that movement between sow pens or stalls and boar pens be straightforward and not time-consuming. If such movement imposes difficulties to managers and stock persons, then, in busy periods, both the efficiency of heat detection and service management are likely to suffer. Thus, in housing design, considerable thought should be given to ease of movement between sow and boar accommodation in order to ease the workload of stockmen and to increase the certainty that heat detection and service management will always be maintained at a high level of efficiency.

House temperature

Cold accommodation at weaning imposes an energy drain on sows and this is accentuated by poor bodily condition at weaning, individual housing and low feed levels.

If sows are going to reproduce normally in cold accommodation, the feed level must be increased as indicated in Table 10·5.

TABLE 10·5. The effects of temperature of sow house on food requirements

	Optimum temperature (unbedded stalls)	*Additional feed required for each 5°C drop in temperature*
Thin sows	22°C	300 g per day
Sows in moderate condition	21°C	170 g per day

Source: Holmes, C. W., Massey University, New Zealand. Close, W. H., Institute of Animal Physiology, Babraham, Cambridge.

This shows that higher feed levels are required by thin sows than those in moderate condition to counteract the effects of cold housing.

It has also been noted above that the stimulus from the boar for sows to start breeding activity is likely to be greater in a warmer environment.

Excessive heat can also adversely effect breeding performance of both sows and boars.

Temperatures of 23 to 27°C in boar and dry sow accommodation can have a very adverse effect on fertility. This is partly through the adverse effect of high temperature on the initial development of the sperms. The sperms take about four weeks to mature so that one can expect problems from services taking place about one month after a hot spell of weather. Often, services carried out at this time are infertile because of defective sperms and, as a result, return rates are high.

To avoid such a problem, the cooling of boars is desirable either by well-insulated buildings protected from the sun, by forced draughts, or by provision of a wallow or sprinkler device. Another factor which may be associated with infertility in very warm weather is the libido and stamina of boars. Since there may be up to a 10°C difference between maximum and minimum temperature

Plate No. **39**

Communication between newly weaned sows and the stimulating boar in an adjacent pen.

in any 24-hour period in a pig building, during hot spells of weather it may be wise to consider concentrating heat detection/service periods in the cooler parts of the day, such as 0700 hrs and 1900 hrs, and avoid the period of maximum temperature around midday.

High temperatures can also have an adverse effect on sows in early pregnancy, embryonic losses being increased.

There is an obvious need, therefore, to prevent both boars and sows being subjected to excessively high temperatures.

Service area

It is vitally important to have a good service area, usually in the boar pen, where both the sow and the boar can obtain a good foothold. Slippery floors can shatter the confidence of both boars and sows. Sows which slip badly when being mounted by the boar may refuse to stand again during that heat period and so twenty-one vital days are lost before the sow has her next heat.

3. SEASONAL EFFECTS

Much evidence is available from the UK, France the other

Western European countries to indicate that lowest fertility prevails in sows weaned from June to September. The problem is one of anoestrus with delays in sows coming in heat and there may also be poorer conception rates. The exact reasons for this seasonal infertility are not known. It may be brought about by decreasing day length or high environmental temperature. It is possible that the problems may be partly an indirect effect of high summer temperatures in reducing food intakes in lactation and resulting in poorer body condition at weaning. Certainly some producers in the UK report a higher incidence of seasonal infertility problems in hotter summers. A well-insulated house, good air flow and the avoidance of excessive stocking all help to keep sows and boars cooler in hot weather. In loose housing with solid-floor mucking, sows and boars will often 'wallow' in this area in hot weather in an attempt to keep cool.

4. THE BOAR

All too often the boar is treated as an unwanted extra instead of a main component of production. How often do we find the boar stuck away in cold, damp, uncomfortable quarters grossly affected with mange and looking generally unhappy? The boar has much greater influence on herd performance than the individual sow and should be treated accordingly. Boars may be afflicted with a number of problems.

STERILITY

Primary sterility

Some boars may fail to produce semen or produce semen which is abnormal in some way. They show normal interest in sows in heat, mount and serve adequately but leave no sows pregnant. Records will soon detect these boars and an examination of semen by a veterinary surgeon will confirm the diagnosis. The cause may be either inherited or developmental. Such affected boars should be slaughtered forthwith. Testicular hypoplasia and testicular degeneration both result in gradual infertility. The former may be inherited while the cause of the latter is not known. The semen produced is abnormal and degenerative changes or disparity in size may be detected in the testes. Only careful scrutiny of the records will bring such cases to light. The diagnosis can only be made by a veterinary surgeon and there is no treatment for either condition.

Secondary sterility
Temporary or permanent sterility may occur after a boar has contracted a febrile disease affecting the body as a whole or after a localised infection in the testes (orchitis). Febrile conditions usually lead to temporary sterility. Poor quality semen may be produced for a period of two to three weeks after infection (e.g., acute erysipelas, acute pneumonia). Sudden high temperatures even in temperate climates can have a similar effect. It is only the young developing sperms which are affected. Since these sperms take approximately four weeks to develop and mature, there will be a gap of four weeks between precipitating disease or high environmental temperature and the boar becoming infertile. This is an important point to remember when checking records of services in relation to the production of small litters.

Treatment
Rest the boar for a month and then semen-test him. Localised disease of the testes usually results in the boar gradually becoming infertile. There may be associated clinical signs such as enlargement, pain and tenderness of the testicles. Pus and pus cells may be detected in the semen. Treatment of such cases may be prolonged, uneconomical and unsuccessful. Boars may also become infertile due to over-use. This may occur as a sequel to infertility in the sows which may become repeat breeders. This is known as the domino effect. Over-use may also occur when one or two boars are preferentially used by the stockman because of their active ability to serve sows quickly and carefully.

ANATOMICAL DISORDERS

Persistence of the frenulum is a condition now being encountered more frequently. In the young immature animal the penis is normally attached to the prepuce by a piece of elastic tissue known as the frenulum. This attachment is released with normal development but occasionally it persists so preventing extension of the erect penis from the sheath.

Treatment Surgical separation—consult a veterinary surgeon.

A small penis improperly erected is another condition occasionally seen. The boar cannot direct or insert the limp penis which often becomes injured. Boars affected with this disorder should be slaughtered. Another condition in this category is coiling of the erect penis within the diverticulum (a sac leading from the prepuce). Ejaculation takes place in the diverticulum and semen will often be seen around the belly of the boar and on the sow's hindquarters. This condition is probably a vice and can only be detected by manual examination of the parts at service.

Treatment
Surgical excision of the diverticulum by a veterinary surgeon.

LACK OF LIBIDO (POOR SEX DRIVE)

This condition is often temporary and may be due in young boars in some instances to a previous upsetting experience such as intimidation by a group of sows when first used or bullying by a larger more dominant male. Lack of a chance to develop normal male habits such as mounting, erection and general male developmental activity may also contribute to the problem. Such boars should be assisted with at least one service using a small sow in full heat.

Genetic causes have also been implicated with lack of libido, especially in Landrace boars. These slow-serving boars are a nuisance, especially in large herds, and often lead to the over-use of another more active boar. Sudden exposure to heat (even in temperate climates) or febrile conditions will also lead to lack of libido. The cure is to rest the animal for two to three weeks in this instance and house in cooler conditions.

Oestrogenic substances (female hormones) produced by certain fungi contaminating foodstuffs can also lead to poor sex drive. In some cases, boars with lack of libido may benefit from a course of hormonal injections.

LOCOMOTOR DISTURBANCES

Failure to mate is often due to some crippling condition affecting the feet, legs or back of the boar. Laminitis, feet sores and cracks and arthritis of the joints will render the boar incapable or unwilling to mount. Injuries to the muscle or ligaments of the back are common. Spondylosing arthritis is a common condition affecting the spinal vertebrae which leads to a gradual seizing-up of the spine and an inability to mount. Occasionally, the floors are so slippery that the boar loses confidence and becomes apprehensive of mounting. 'Skating rink' conditions are all too common in badly-designed pens.

MISCELLANEOUS PROBLEMS

Bleeding after service
Bleeding, sometimes copious, may occur just after service. The blood may come from the prepuce, the body of the penis or from within the urogenital tract. Sows often return to service in this case. The condition should be taken seriously and veterinary advice sought immediately.

Anal service
Young boars may occasionally start servicing into the rectum. This may become a habit unless checked. Not only does it lead to gross infertility but also to damage of the rectum. Services, especially in young boars, should be observed carefully. In many cases *visual* examination only will not detect this habit. It is essential to slide the hand gently upwards towards the vulva where the body of the penis should be felt during normal copulation.

Service technique
Service technique is a very important factor and this can differ markedly between boars.

By service technique is meant the time taken by a boar to mount a sow and how settled he is during service. Boars which take a long time to mount try the patience of the stockmen, while boars which continually paddle with their front feet during service, chafing the shoulders of the sow in the process, severely tax the patience of the sow. The result is that often the sow runs away before proper ejaculation gets under way and often refuses to stand again for any other boar during that heat period, and so a valuable twenty-one days is lost. Boars differ markedly in service 'technique' and it is a good idea to see a boar serving a sow and be satisfied about his technique before he is purchased.

GENETIC DIFFERENCES

An apparent difference in fertility (number born) between boars of different breeds was indicated from an analysis of the records of a large commercial herd. This information is presented in Table 10·6.

TABLE 10·6. Relative litter sizes out of Large White and Landrace Boars

	Breed of sire	
	Large White (LW)	*Landrace (LR)*
Number of sires	22	12
Number of litters	445	195
Mean per litter (alive + stillborn)	11·01	10·09

The sows in this herd were almost entirely first crosses, either LR × LW or LW × LR, and almost one extra pig resulted from services by Large White boars.

From analysis of records in another large herd, apparent differences between sire breeds emerged in terms of conception rate. The relevant figures are presented in Table 10·7.

TABLE 10·7. Percentage service returns to Large White and Landrace Boars

	Breed of sire	
	Large White	*Landrace*
Number sows served	408	507
Returns to first service (per cent)	9·8	17·6

Thus, return rates were almost twice as high to Landrace than to Large White boars. In this herd, Landrace boars were mated to Large White sows and vice versa, and so the difference in conception rate emerging could be due, at least in part, to the breed of sow involved. However, this latter possibility is very unlikely and it would appear that it is differences in fertility between the breeds of sire which are being demonstrated.

These apparent differences in fertility between breeds of sire are likely to be due to differences in quality of semen and/or libido or stamina of boars. It may well be that, in order to achieve comparable results, on average, from Landrace and Large White boars in subsequent conception rates and litter size, Landrace boars should be used more sparingly than Large White boars since the Landrace boar may be more subject to the effects of overwork in depressing fertility than the Large White.

MONITORING BOAR FERTILITY

Boars differ in fertility and the fertility of any one boar will vary over a period of time. It would prove very useful if boars could be classified on the basis of fertility and this would assist management in deciding on the relative use to be made of different boars. However, since it would appear that fertility can vary quite markedly over short periods of time, a classification of boars made at any specific time is unlikely to be very reliable for very long. Mature boars should manage between three and four double services per week, on average, but all producers have some boars which can be worked harder and still produce good results and others which cannot be used so frequently without adverse breeding results occurring. It is important to keep service records on each individual boar up to date so that any fertility problems associated with the boar can be spotted promptly and appropriate action taken.

USEFUL GUIDELINES—A SUMMARY

- House the boar comfortably adjacent to newly-weaned sows.
- Protect from extremes of temperature.
- Ration so as to keep in fit but not fat condition.
- Wait until the boar is 7–8 months old before he is used.
- Allow four services a week until a year old and then six services a week thereafter. Double this number of services in a week appears to create no problems provided the boar is rested for 10–14 days afterwards.
- In very hot weather, use during the cooler periods of the day.
- Do not rest for longer than thirty days.
- Serve gilts or sows in the boar's pen.
- Ensure floor conditions are good.
- Observe services closely (examine by hand if necessary).
- Keep good records of boar health, use, treatment and performance.

5. AGE AT WEANING

When litters are weaned before ten days of age, there can often be a considerable delay of up to three weeks before sows come into oestrus: also, there can be a wide range in timing of appearance of heat after weaning. While sows weaned between ten and fifteen days following farrowing may take slightly longer to come in heat than those weaned at a later stage, the increase in the weaning-to-oestrus interval is certainly not large under good management (see Chapter 13, Table 13·2).

While reasonably short weaning-to-mating intervals can be achieved following early weaning, litter size following early weaning tends to be depressed as indicated in Table 10·8.

TABLE 10·8. Litter size in relation to length of previous lactation

	Length of previous lactation (days)			
	16–20	*21–25*	*26–30*	*31–35*
Number litters	167	182	187	207
Average litter size	10·1	10·5	10·7	11·3

Source: Large commercial unit.

Thus, between 16 and 35 days, for each delay in weaning of five days, litter size tended to increase by about 0·4 of a piglet on this unit.

In other studies, the increase in litter size has been about 0·2 piglet for each delay of five days when weaning between two and five weeks of age.

Work at the University of Nottingham has indicated that this depression in litter size with earlier weaning is caused not by reduced ovulation rate but by increased losses of embryos (see Table 10·9).

TABLE 10·9. Relationship of age at weaning to embryo losses

	Age at weaning (days)		
	7	*21*	*42*
Number of:			
Ova shed	15·6	16·8	16·9
Embryos at 20 days	9·2	11·5	13·4
Embryos lost	6·4	5·3	3·5

Source: Varley, M. A. and Cole, D. J. A. (1975). University of Nottingham.

The reason for the greater losses of embryos associated with early weaning is thought to be connected with the state of the uterus and the time taken for the whole reproductive system to be restored to full and efficient functioning following farrowing. The uterus takes a certain minimum time to be restored to its fully functional state after farrowing and the sooner service and conception take place following farrowing, the less well prepared will the uterus be to accept and nourish all embryos.

Consequently, earlier weaning is associated with depressed litter size and, if very early (before fifteen days), there is a slightly longer delay from weaning to mating.

6. LITTER NUMBER

It is well established that sows are most difficult to breed following their first lactation. This is illustrated in Figure 10·3 which is based on data from a French survey covering many farms and some 150,000 litters.

The reason for the greater delay in breeding activity following weaning the first litter is likely to be associated partly with the tendency on many farms for some gilts to lose a good deal of condition during lactation and be fairly thin at weaning. It is also likely to be partly linked to the fact that after weaning her first litter the young sow is still growing actively and there may be more competition for nutrients as between growth and reproduction in such animals than in more mature sows. Even with early weaning, gilts tend to lose more weight (in relative terms) than sows while suckling. The end result of lactation in the young growing animal is

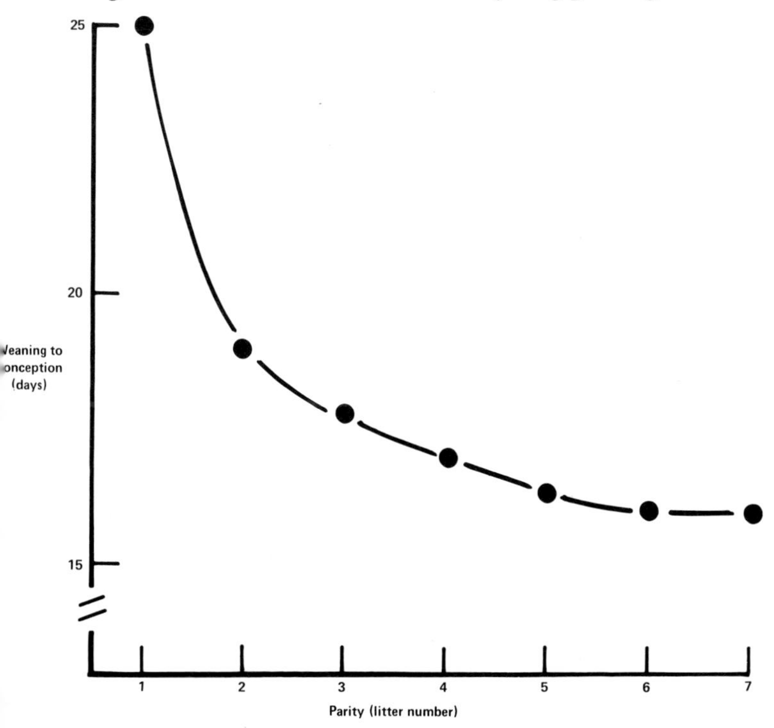

Figure 10·3. Weaning-to-conception interval in relation to parity.

Source: Aumaitre, A., Dagorn, J., Legault, C. & Le Denmat, M. (1976). French Survey Results.

a severe shock to the body systems. Nature protects the weak by shutting down the breeding mechanism until such time as the animal is fit to breed again.

One should attempt to avoid poor condition in gilts when their first litter is weaned by increasing food intake while suckling. If there is a low appetite problem in lactation, this can often be traced to excessive feed intake in the first pregnancy. Thus, reduced intakes in pregnancy will result in less fat gilts at farrowing, better appetite in lactation and help to avoid poor bodily condition at weaning.

7. GENOTYPE

It is well established that crossbred sows are more efficient breeders than the mean of the constituent purebreds. Reproductive traits such as age at puberty, breeding regularity, embryo survival and piglet viability are those most likely to receive a boost from hybrid vigour.

8. GENITAL AND LOCOMOTOR DISORDERS

Normal mating may be prevented by several disorders—especially those of the locomotor system or genital system.

*Intersexuality**

Animals possessing genes, gonads (testes or ovaries) or accessory genital organs of both types are classed as intersexes.

Up to 0·5 per cent of pigs may be affected. When both kinds of gonads are present (internally) the pig appears outwardly to be a female (true hermaphrodite). (These animals may even come in heat and conceive but the resultant litter size is usually small. Such cases are the exception to the rule.) Male pseudo-hermaphrodites are relatively common in pigs. These have external female characteristics with enlarged clitoris, vestigial prepuce and incomplete vagina. At puberty such animals adopt masculine type of behaviour. These animals have an internal uterus and testes but are *genetically* female. Careful examination of gilts at selection should detect such cases.

Several anatomical abnormalities of the female reproductive tract have been demonstrated. These may prevent entry of the penis or prevent locking at service.

*A. E. Wrathall, MAFF, Weybridge, Surrey

Locomotor disorders

Disorders of the locomotor system lead to complete or partial inability of the sow to accept the boar. Osteo-arthritis of the limbs or spine, fusion of the spinal column, epiphysiolysis, apophysiolysis and lameness are examples. These conditions (lameness excepted) can only be diagnosed by a veterinary surgeon. Failure to mate is often entirely due to wet, slippery floors which may be either solid or slotted. Animals confined on such floors quickly lose confidence and are usually injured.

9. DISEASE IN RELATION TO EMBRYONIC LOSS AND ABORTION

Approximately 35 per cent of fertilised ova are lost in normal pigs. When death of the embryo occurs during the pre-attachment phase, i.e. the first 14 days, they are reabsorbed by the dam and the sow will return to heat in 19–21 days after service. When death of all or some of the embryos occurs during the embryonic phase, i.e. between attachment at Day 14 up to Day 35, the sow may abort, especially if all are dead or she may reabsorb the embryos as they gradually die off. Abortion at this stage is characterised by an insignificant discharge which may be blood-stained. In this case, there will be a delayed return to oestrus. When only a proportion of the embryos have died, the sow may maintain pregnancy. Most of the dead embryos will be completely reabsorbed but some mummified remnants may still be seen at birth. After Day 35 the foetuses are recognisable as small pigs. Partial litter loss after this stage is usually indicated by the presence of mummified or partially reabsorbed foetuses at birth. Death of all or nearly all the litter results in abortion. There is a risk that pregnancy will not be maintained unless there are four or more viable foetuses in the womb.

Prenatal loss may be divided into two broad categories:

- Those caused by failure of the dam (maternal failure leading to abortion or complete absorption).
- Those caused by disease of the foetus (foetal failure leading to reabsorption or mummification or abortion).

Occasionally both types of failure may be present together, caused by overcrowding in the womb and the presence of a slow-acting virus. Maternal failure results in either abortion (normal parturition at the wrong time) or complete reabsorption in the early stages. Most of the latter will return to heat, but some which have previously been ill or in poor condition revert to anoestrus. Aborted foetuses due to maternal failure are usually fresh looking, exhibit no unusual body size variation and show no signs of disease. The most common causes of maternal failure are infection leading to septicaemia, bacteraemia, toxaemia and fever. These lead to cessation of the hormonal control of pregnancy. Stress may also act more insidiously over a long period and also by affecting hormonal control of pregnancy.

Infectious disorders of reproduction have been classified into three groups:

Group I infections

These are caused by organisms commonly found in and around the pigs, which exist together for mutual benefit (commensals). The majority are bacteria but viruses, mycoplasma and fungi are also involved. These organisms only become active and produce disease when maternal resistance has been lowered. Examples of such organisms are erysipelas, salmonella and *E. coli*. Some of these organisms may already be present in normal wombs (see Chapter 3). These organisms may also produce an effect by multiplying in the gut, the skin, the whole body or in the womb itself. Outbreaks of sporadic disease are to be expected when these organisms are involved. Group I infections which enter the womb may cause death of a few foetuses or of all the foetuses. In early gestation, this will lead to complete absorption but in later gestation this will either give rise to abortion of diseased foetuses or the birth of both mummified and diseased foetuses in the presence of live-born healthy piglets.

Control of such infections is not easy and is never likely to be wholly achieved. Nevertheless, some control may be achieved by:

- Reducing the challenge of infection.
- Promoting resistance of susceptible animals.
- Removing or avoiding predisposing factors.

Sows or gilts which are stressed by a combination of overcrowding, excessive fighting, poor bodily condition and housed in a cold, draughty environment with bad floors are likely to have lower resistance and are more susceptible to Group I infections. Body sores are frequently seen in thin sows housed in stalls, and these sores become a focus of infection which either leads to toxaemia or bacteraemia and a consequent cessation of hormonal control of pregnancy. Abortions are frequently seen in long, cold winters when this condition becomes all too common. Total resistance can be improved by ensuring that sows are in good bodily condition and housed in warm, comfortable quarters. Specific resistance to some diseases such as erysipelas and *E. coli* can also be achieved by the use of vaccines. Intercurrent infestation with lice, mange and worms should also be dealt with.

The weight of environmental infection can also be reduced by taking measures such as the periodic disinfection and vacation of buildings. When sows are thought to be harbouring potentially dangerous organisms in the vagina, the genital tract should be washed out with a suitable antibiotic preparation prior to service.

Group II infections

These are caused by certain common contagious micro-organisms (viruses in this case) which are common in a high proportion of pig herds, to which pigs normally develop a strong immunity just after weaning. The common agents in this group are porcine enteroviruses and parvoviruses—also referred to as SMEDI viruses [stillbirth (S), mummification (M), embryonic death (ED) and infertility (I)]. There are at least ten enterovirus groups and one group of parvoviruses, both with sub-strains within each group. These viruses are commonly found in the gut of normal sows, and the majority of herds producing weaner pigs pick these up and develop a strong immunity. However, when exposure to infection is delayed until puberty or later, the consequences may be grave depending on the strain of virus or number of viruses involved.

In the pregnant gilt or sow which has not been previously exposed, the virus can cross the placental barrier and cause death of the foetuses. The earlier in pregnancy the infection occurs, the more serious the results. Infection of the embryos during the first four to five weeks of pregnancy usually results in the death of the litter followed by complete reabsorption and delayed return to heat or sometimes anoestrus. This anoestrous condition may prevail throughout the gestation period giving rise to false pregnancies. Such animals go through the early signs of farrowing with milk production but produce no foetuses.

When infection occurs later in pregnancy, the foetuses may die and will be born as mummified foetuses or as weakly-born diseased foetuses. Sometimes, litters will be produced containing mummified foetuses, weakly live-born piglets and healthy live-born piglets at the same time. Some strains of virus are less pathogenic and infect the foetuses without producing disease.

Therefore, the presence of virus in diseased or aborted foetuses must not be taken as the cause of the incident. Abortion is not a feature of SMEDI virus infection. The usual outcome of whole litter death at the two- to three-month stage is mummification. Litters of mummified pigs may be carried for many weeks after the normal parturition date has passed.

Outbreaks of SMEDI may be expected when susceptible gilts (usually) are purchased and mixed with the parent herd at service or are exposed to the virus for the first time during pregnancy. On the other hand, a fresh strain of virus may be introduced to an already stabilised herd via contaminated boots or a newly-purchased boar. The diagnosis is difficult mainly because of the ubiquity of these viruses and the fact that the virus may be found in normal healthy foetuses. A specific diagnosis can only be made by a veterinary surgeon. All diseased or aborted piglets and the placenta from each litter should be kept for examination by a veterinary surgeon.

Control measures against Group II infections are difficult to apply and complex in nature. Modern intensive pig-keeping with emphasis on hygiene, quarantine and prevention of disease spread has created herds of low immune status with regard to the enteroviruses and parvoviruses. Frequent purchase of tested boars and gilts has led to exposure of these animals at the wrong time. There is no one control solution which will satisfactorily cover both Group I and Group II infections.

Generally speaking, efforts should be made to expose all breeding stock to the microflora of the herd up to three weeks before service, after which time they should not be exposed until after farrowing. However, there are many herds where continuous exposure before puberty and throughout gestation until farrowing has been successsful. In such cases, no fresh introductions of viruses occurred and such a situation is quite foreseeable in small herds (e.g. those under 150 sows).

Another compromise in herds of any size is to expose gilts three weeks before service and for the last three weeks of pregnancy. This ensures good immunity to Group I infections, while Group II infections will have much less effect on the foetus when introduced in late gestation. An excellent rule is never buy or sell *pregnant* gilts and buy gilts from one source if possible. Exposure may be carried out in strawed yards by putting in liberal quantities of dung from older stock or introducing non-pregnant stock (except for those within two to three weeks of mating). Once an outbreak has begun, it is too late to stop it. However, measures should be taken to infect all pigs (except pregnant animals) by distribution of faeces, afterbirth and any available foetal material.

Following such exposure, animals will develop immunity to the infectious agent or agents. The exception is the maiden gilt which may still carry maternal antibodies. Such animals will not become immune by natural exposure to the virus but will probably become susceptible to infection during the first pregnancy. If such animals escape exposure during the first pregnancy they are likely to become infected during the second and will exhibit the typical signs of SMEDI infection.

A SMEDI vaccine is available in some countries but reports suggest it is little better than natural controlled exposure to the virus. Once immunity is acquired it lasts for the lifetime of the sow. According to Professor Leman of the University of Minnesota, an infected mummified foetus is equivalent to 500 vaccine doses.

Minimal disease herds are unlikely to be free of SMEDI viruses and if they are, the probability that they will become infected is high. Control and preventive measures of Group II infections in MD herds are the same as for conventional herds.

Group III infections

Group III infections are specific infectious diseases of reproduction, e.g., *Brucella suis* (absent from the UK), leptospiral infections, toxoplasmosis, congenital tremor Type 11A, swine fever and Aujeszky's disease are examples of others. Congenital Tremor Type 11A and Aujeszky's disease are described in Chapter 3 in this book, and the others are rarely diagnosed in the UK.

Diagnosis of Group III infections can only be made by a veterinary surgeon, and advice on control measures will vary according to the organism responsible.

Miscellaneous factors

Prenatal loss may also be caused by specific vitamin deficiencies and certain toxic or teratogenic agents.

Vitamin A, which is derived from plants, is essential for the growth and development of the foetus. Deficiency in the foetus will lead to the birth of dead, weak and malformed piglets. Absence of eyes, very small eyes (microphthalmia), cleft palate and sometimes hydrocephalus are signs of congenital vitamin A deficiency. There may be no evidence of such a deficiency in the dam. Should the dam suddenly acquire a normal intake of vitamin

A late in gestation, the newborn affected piglets will probably have normal levels of this vitamin in the tissues. The malformations mentioned may also be confused with some genetic diseases. A veterinary surgeon and nutritionist should be consulted if piglets are born with the above-mentioned abnormalities. Finally, vitamin A deficiency does not cause abortion in sows.

Sows which have consumed hemlock and/or jimson weed will produce malformed piglets. The affected piglets have a characteristic twisting and bending of the limbs (arthrogryposis). Ingestion of coal tar, toxins from certain moulds and ingestion of various drugs will also lead to foetal death and abnormality. Fortunately, sows rarely have the chance of consuming such substances.

Problems at farrowing have been covered elsewhere in this book.

In summary, it may be said that reproductive problems are often complex, costly, insidious in nature and difficult to diagnose precisely. Thankfully, a precise diagnosis of the aetiology is not always necessary. The epidemiology, history and pattern of an outbreak of infertility, together with an examination of well-kept records, are often all that is necessary for a veterinary surgeon to give sound practical advice on control and preventive measures.

Use of hormones—words of caution

When a breeding problem arises or when such is suspected, use of hormones in an attempt to remedy such should be used only when all else has failed. Widespread use of hormone injections is an indication of management failure. Hormones should be used under veterinary supervision as use of the wrong hormone may have disastrous results.

DETECTING SHORTFALLS IN BREEDING PERFORMANCE

Recording

In a herd of sows, it is essential to monitor performance of each sow regularly with a view to detecting shortfalls and problems promptly and taking appropriate and timely remedial action. An essential basis for such recording and monitoring is a good identification system for sows and litters (see Plates 41A and 41B).

Preferably, the sow should have a single record card which should follow her from dry sow pens to farrowing house to service pens, back to the dry sow unit, and so on. This card may be of the type shown in Chapter 14, Figure 14·1, and gives all relevant information on background (sire and dam and genotype of these), date of birth, date at all services, farrowing dates, numbers born dead, alive and weaned, weaning weights, details of fostering, number and estimated causes of death, disease problems in the sow and piglets and details of veterinary treatment.

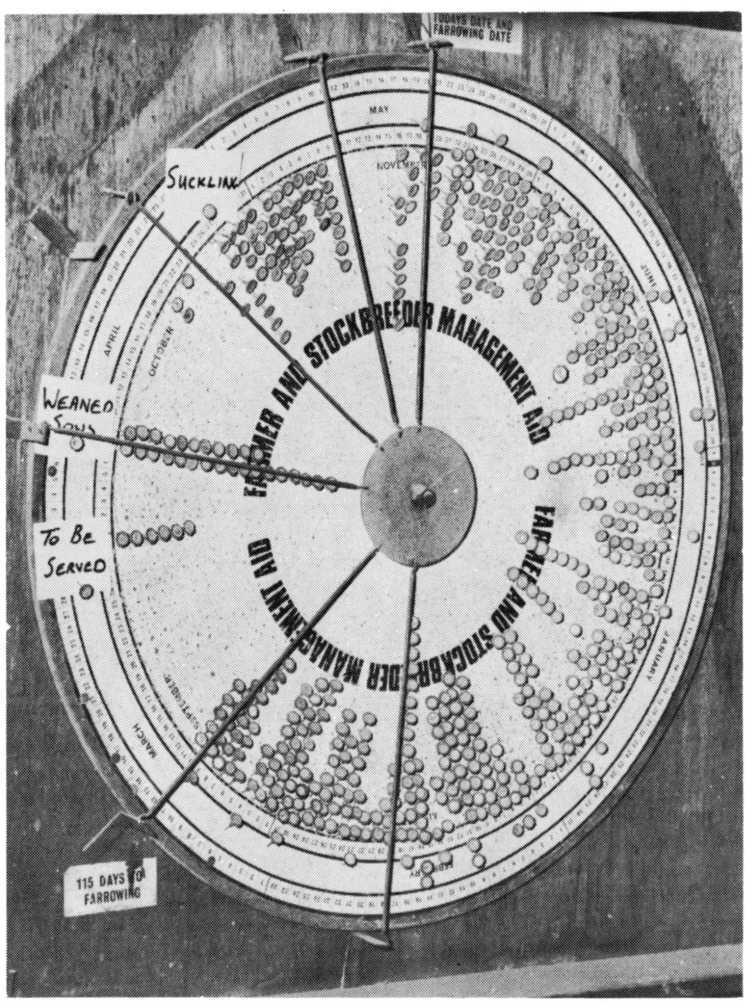

Plate No. **40**

A herd indicator board with details of each sow in the herd, and the particular stage each is at, is a very useful management aid.

There should be a herd indicator board (see Plate 40) on which all sows in the herd are represented by corresponding indicator pins, placed at the appropriate stage of the reproductive cycle. Newly-weaned sows should be kept as a tightly-knit unit by being housed adjacent to each other, and their indicator pins should be clearly identified and kept together on the herd indicator board. As they are served, their indicator pins can be moved away from those of their unserved contemporaries. This will help further to feature the unserved sows in the minds of the manager and stockmen so that they command the required amount of attention. Sows approaching their 21-day return date should also be clearly identified on the indicator board so as to provide constant reminders to stockmen to check for service returns.

It is extremely useful and helpful to keep an up-to-date record of

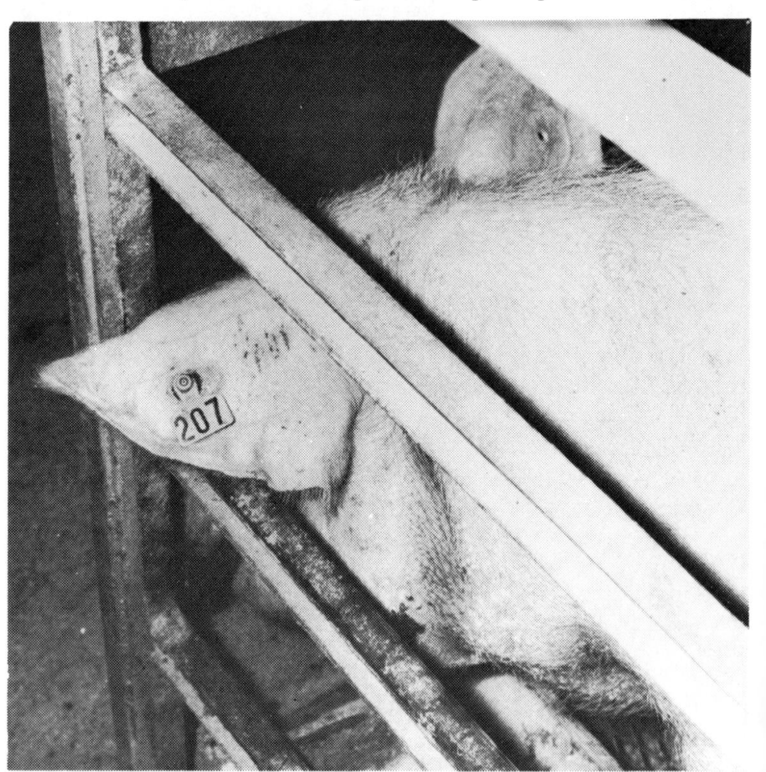

Plate No. **41A**

It is very important that sows be readily identified by using numbered tags or a similar system which can be easily read.

Plate No. **41B**

Each sow should also have a card containing details of performance which should hang adjacent to her pen and be moved along with her.

each boar in the herd, recording all services, returns, abortions and litter details so that his breeding performance can be readily checked.

Piglets within a litter can be given a single identification mark so that the litter performance details at weaning and slaughter can be used to guide selection of replacement stock in those herds breeding their own replacements.

All of these sow, boar and litter records are also, of course, an essential basis for deciding on culling policy.

Extracting breeding performance data from individual sow and boar cards and summarising this weekly in graphic charts is an extremely useful way of keeping a constant check on all aspects of breeding performance. These also help to detect the start of a problem promptly and often allow corrective action to be taken before the problem becomes too severe. Such weekly graphic charts may feature number of maiden gilts in the herd, number of gilts mated, number of sows weaned, number of sows mated, days weaning to service, returns to service, abortions, numbers born mummified, stillborn, weakly liveborn and normal liveborn, farrowing difficulties, incidence of post-farrowing disease in sows and neonatal disease in piglets.

PREGNANCY DIAGNOSIS

After sows have been served and they are checked for 'returns' nineteen to twenty-three days later, if they show no sign of oestrus at this stage they are assumed to be 'in pig' and may receive scant attention afterwards in terms either of checking for heat or of confirming pregnancy. Prior to the development of pregnancy diagnosis techniques, often the stage following the twenty-one day return date, when sows were next subjected to close scrutiny, was about seven days prior to the due-to-farrow date when they were about to be moved to farrowing accommodation. Some at this stage may have been found to be not in pig. This meant an animal which was wholly unproductive for almost four months which was using up expensive feed and other inputs and which was occupying valuable accommodation for no return.

Pregnancy diagnosis carried out from about three weeks after service is proving a very useful technique in isolating the unproductive animal at an early stage so that she can be stimulated to rebreed or be culled as the occasion demands.

A variety of techniques are available for pregnancy diagnosis and these have been studied in detail by Dr G. W. Dyck of the Agricultural Research Station, Brandon, Manitoba, Canada. The main methods available are as follows:

Vaginal biopsy

This is based on taking a very small sample of lining of the vagina and having the number of cell layers counted. The test can be carried out from about eighteen days after service up to term. Normally, results are available two to three days after sampling. The test confirms pregnancy with 90 to 100 per cent accuracy. On the other hand, confirmation of barren animals is only about 75 per cent accurate, on average.

Blood progesterone level

This involves taking a blood sample seventeen to twenty-four days after service and assessing plasma progesterone concentration. Chemical analysis usually takes two to three days to complete. The test confirms pregnancy with 92 to 97 per cent accuracy. Confirmation of barren animals can be 95 per cent accurate. However, the occasional pregnant sow with a low progesterone concentration may be called non-pregnant while non-pregnant sows with extended or short oestrous cycles may be diagnosed as pregnant.

Blood oestrogens
High concentrations of oestrone and oestrone sulphate are produced by the developing foetuses and therefore the blood levels can provide an indication of pregnancy. The test can be carried out between twenty days and term but best results are obtainable by blood sampling twenty to thirty days after service. Estimation of blood oestrone or oestrone sulphate levels at this stage can confirm either pregnancy or barrenness with over 95 per cent accuracy. The occasional animal with a low oestrogen concentration may turn out to be pregnant. Animals in oestrus may have sufficiently high oestrogen concentrations to be incorrectly diagnosed as pregnant. For this reason, blood samples are not normally collected until after the expected time of oestrus.

Manual method
This method involves examination of various parts of the reproductive tract per rectum from thirty days after service onwards. The main aspects to examine are the size of the middle uterine artery and the pulse of the blood flow through the artery. On the farm, it is essential that the examination is carried out by a veterinary surgeon. With an experienced and skilful operator, pregnancy can be confirmed with 80 to 100 per cent accuracy, while confirmation of barrenness is usually only about 50 per cent accurate.

Measurement of foetal pulse
The foetal pulse can be detected by reflected sound waves (Doppler Principle), these being magnified and monitored by auditory means on devices such as the veterinary fetometer. The test can be carried out from about twenty-five days after service onwards. Experience is required to recognise the foetal pulse and errors in the alignment of the transducer may produce errors in pregnancy detection. A skilful and experienced operator can be 95 to 99 per cent accurate both in confirming pregnancy and barrenness. However, a second examination at mid-pregnancy on sows which were in the doubtful category at the first examination is necessary.

Detection of fluid-filled uterus
The fluid-filled uterus characteristic of pregnancy can be detected by reflected sound waves from around the thirtieth day of pregnancy. A wide variety of ultrasonic machines are available for this purpose and some of these can also be used for measuring backfat thickness. Errors in the alignment of the transducer may

produce errors in pregnancy detection, and a full bladder, with the wrong alignment, can give a false positive result. In addition, pregnant animals with low foetal fluid volumes may not be confirmed as being in-pig. Therefore, a second examination at mid-pregnancy of sows which were in the doubtful category at the first examination is necessary.

Therefore, these various techniques can be used to confirm pregnancy or detect barren animals from three to four weeks after service onwards. In diagnosis, pregnancy can be wrongly confirmed and barrennes can be wrongly diagnosed. Both types of mistake are equally serious and emphasise the need for skilled operators of the equipment. However, a positive diagnosis of pregnancy almost invariably turns out to be correct, whereas a negative finding and those in the doubtful category should be rechecked one to three weeks later. Of course, some sows can abort or reabsorb their conceptus after positive pregnancy diagnosis and, if such abortion is not noticed at the time, these sows will some time later be found to be barren.

Pregnancy diagnosis is of particular value in herds with infertility problems, particularly those of anoestrus following first service.

Pregnancy diagnosis equipment should be light, easily read, easily carried, be robust and it is preferable that they are powered by rechargeable batteries.

ENSURING GOOD BREEDING PERFORMANCE—CHECK LISTS

Having outlined the main factors controlling sow reproduction and ways of manipulating these in practice, we are now in a position to summarise the major points by providing check lists in relation to ensuring:

(1) Prompt post-weaning oestrus.
(2) High oestrus detection rates.
(3) Fertile service.
(4) High numbers born.

A general requisite for achieving all four objectives is the maintenance of good health, adequate housing and nutrition. Nutrition in one stage cannot be dissociated from that in another stage of the reproductive cycle, and the general objective in relation to achieving a high level of breeding performance is to ensure that sows gain about 10 to 15 kg liveweight from one weaning to the next up to the fourth to fifth litter. This necessitates treating sows as individuals and, therefore, providing individual feeding. The more specific requirements for achieving the above objectives are detailed below.

1. ENSURING PROMPT POST-WEANING HEAT

(1) Provide adequate nutrition in lactation, especially with regard to protein quality of diet.
(2) Avoid excessive loss of condition while suckling.
(3) Provide comfort and warmth at weaning.
(2) Feed liberally from weaning to service (a multi-vitamin injection at weaning is often beneficial).
(5) House sows adjacent to boar from weaning.
(6) Put sows into small groups at weaning in comfortable, spacious pens with individual feeders. Alternatively, pen individually in stalls.
(7) Use crossbred sows.
(8) Hormone therapy to stimulate heat should only be used when all else has failed.

2. HEAT DETECTION

(1) Heat is easier to detect in sows grouped in small lots at weaning than in individually-stalled sows.
(2) Inspect newly-weaned sows twice daily.
(3) Take the sows to the boar to check for heat. If this is not possible in the case of dry sow stalls, run the boar up the front of the sows, not behind them.
(4) House the boar next to sows.
(5) If heat is suspected but not confirmed by one boar, try the sow with another boar.

3. HIGH CONCEPTION RATES

(1) Ensure good health.
(2) Check boars for fertility.
(3) Select boars with good service ‘technique’.
(4) Manage boars well (housing, feeding etc).
(5) Protect boars from high environmental temperature.
(6) Keep boars in a cool place in hot weather.

(7) Ensure an adequate, non-slip service area.
(8) Avoid over-and under-use of boars.
(9) Supervise all services.
(10) Avoid mixing sows in the three weeks or so after mating.
(11) At least double-serve sows.

4. HIGH NUMBERS BORN

(1) Ensure good health status.
(2) Use crossbred or hybrid sows.
(3) Avoid mating gilts at first heat.
(4) Flush gilts about ten days before service.
(5) Ensure proper timing of service (at least double service).
(6) Large White boars, on average, leave bigger litters than Landrace.

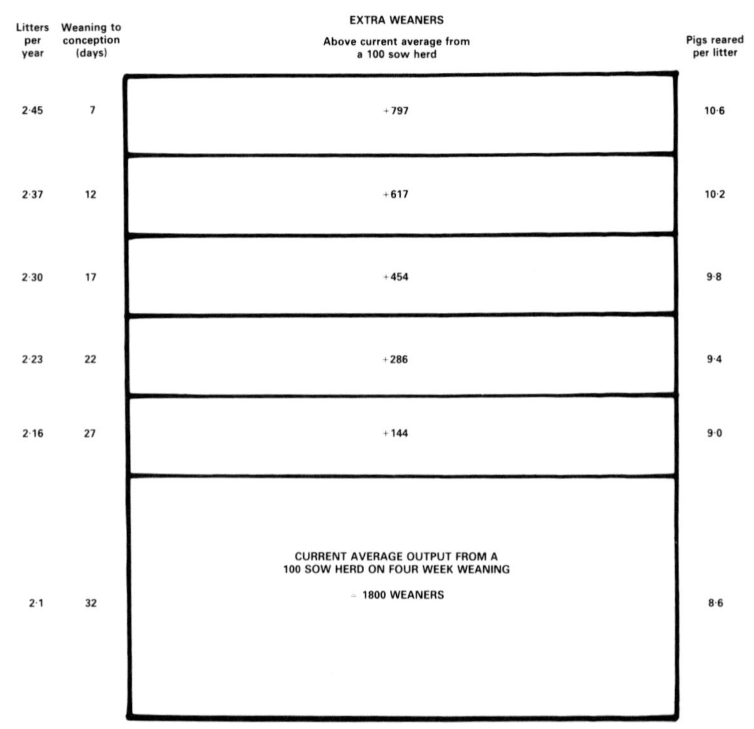

Figure 10·4. The incentive to reduce weaning-to-conception interval and increase numbers born and reared per litter.

(7) Keep sows contented and comfortable in early pregnancy.
(8) Protect sows from high environmental temperature in early pregnancy.
(9) Avoid mixing of sows in early pregnancy.
(10) Provide adequate housing throughout pregnancy.
(11) Provide adequate nutrition, with individual feeding but avoid high-level feeding after service.
(12) Ensure timely culling.
(13) Avoid very early weaning.

The great incentive to pay intimate attention to detail in preparing for mating, service management and pregnancy care is illustrated in Figure 10·4.

CONCLUSION

To ensure that everything is perfect in preparation for, and during, service is absolutely vital. Unless (i) all sows come in heat promptly after weaning, (ii) large numbers of ova are shed, (iii) the maximum number of these are fertilised and (iv) high conception rates achieved, no matter how good the system and management are after this stage, the potential for the maximum number of farrowings and piglets will not be created for later exploitation through good management in pregnancy, farrowing, rearing and thereafter up to slaugher.

Chapter 11

FEEDING THE SOW

No BOOK on the sow would be complete without dealing with its nutrition since food costs constitute about 70 per cent of the total costs in weaner production. Nutrition is a vast and complex subject and readers wishing details of nutrient requirements of the sow and her piglets should refer to 'The Nutrient Requirements of Pigs' published by the Commonwealth Agricultural Bureaux in the UK or to 'The Nutrient Requirements of Swine' published by the National Research Council in the USA. The present chapter aims to summarise the most important practical aspects of sow feeding.

COMPONENTS OF SOW OUTPUT

In evolving an efficient feeding system for sows, one must be aware of the effects of nutrition on the major components of sow output. The most comprehensive measure of sow output, at any specific age at weaning, is the number of healthy weaners per sow per year and this is made up of several components as in Figure 11·1.

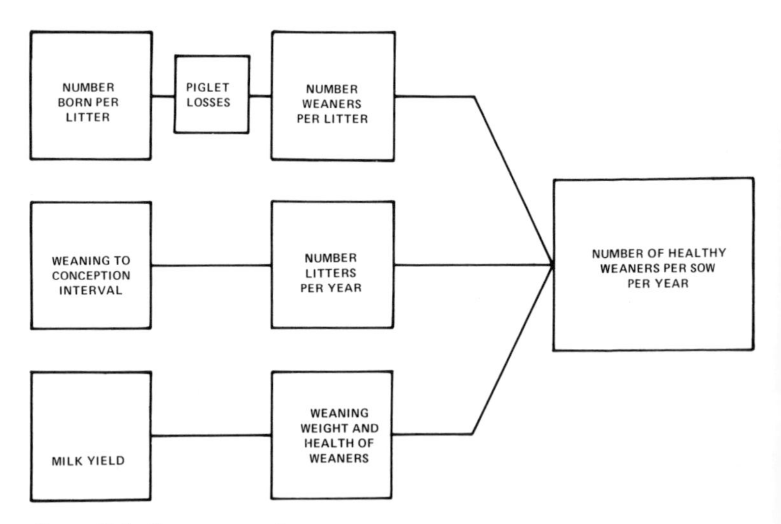

Figure 11·1. Components of sow output.

Thus, one must be aware of the effects of nutrition on numbers born per litter, piglet viability, milk yield and breeding regularity as a basis for evolving a sound feeding programme for sows.

How, then, can feeding be best manipulated in order to put this most expensive of all inputs to most efficient use and maximise the number of sound healthy weaners per sow per year? Before attempting to answer that question, let us outline the major dietary factors and comment on the interdependence of the various stages of the reproductive cycle in relation to feeding policy.

MAJOR DIETARY COMPONENTS

The major components of a diet for sows are:

Energy; protein/amino acids; vitamins; and minerals.

The provisions of vitamins and minerals will not be discussed as this is a very specialised field which is very adequately covered in the publications on nutritional requirements referred to on the previous page. Moreover, all reputable food compounders produce comprehensive vitamin/mineral supplements for sow diets, and if instructions are followed, there is not normally any difficulty in ensuring adequate levels of essential minerals and vitamins in sow diets.

Thus, the nutrients to be considered in this chapter are energy, protein and amino acids, the latter being the constituents of protein. The essential amino acid most likely to be deficient in UK diets is lysine and therefore the levels of this amino acid should be given most attention in the UK.

In considering the effects of energy, this can be expressed either in megajoules (MJ) of digestible energy (DE) per kg of diet or else in terms of the quantity of a diet with a known energy content. A simple diet for sows is detailed in Table 11·1.

Some producers insist on a proportion of white fish meal in a sow diet, while others have been operating very successfully on such a diet as the above. This conventional diet for sows has 14·1 per cent crude protein in the diet as fed, and its energy content can be calculated by allocating an energy value of 12·7 megajoules (MJ) of digestible energy (DE) per kg to barley and of 15·0 MJ DE per kg to soya bean meal. By simple arithmetic, the energy value of the above diet as fed can be calculated to be 12·7 MJ DE per kg of diet.

Regarding the percentage of the amino acid lysine in this diet,

TABLE 11·1. Example of a conventional UK diet for sows

Ingredient	*Per cent composition*	*Per cent crude protein (CP)*	*kg CP in 100 kg diet*
Barley	85·2	10	8·5
Soya bean meal	12·5	45	5.6
Di Calcium phosphate	0·5		
Limestone flour	1·3		
Salt	0·3		
Vitamin/trace mineral supplement	0·2		
	100·0		14.1

with barley having an average lysine content of 0·36 per cent and soya bean meal a lysine content of 2·8 per cent, the diet as fed will have a lysine content of 0·66 per cent or 6·6 g of lysine per kg of diet.

Thus, 2 kg of the above diet will provide the sow with 2 × 12·7 (= 25·4) MJ of digestible energy (DE), 2 × 0·141 (= 0·282 kg) or 282 g of protein and 13·2 g of lysine.

In summary, then, the composition of the diet detailed in Table 1 is as follows:

Energy 12·7 MJ DE per kg.
Protein 14·1 per cent (or 141 g per kg diet).
Lysine 0·66 per cent (or 6·6 g per kg diet).

For the purposes of later discussion this will be termed a conventional UK diet, and energy will be expressed as a given quantity of this diet to put it into more practically meaningful terms.

It is worth emphasising at this stage that reference to protein level of the diet is not very meaningful without reference to the quality of the protein. Certain diets come up to recommended levels in terms of crude protein but can be deficient in lysine or other amino acids.

INTERDEPENDENCE OF VARIOUS STAGES OF THE REPRODUCTIVE CYCLE

The excellent and painstaking work carried out on pig nutrition at various centres in several countries has demonstrated that nutrition in one part of the reproductive cycle of the sow cannot be dissociated from nutrition in another phase. One cannot consider nutrition in pregnancy separately from nutrition in lactation, for nutrition in pregnancy has implications in lactation, while feeding in lactation can influence both the promptness of rebreeding after weaning and events in the subsequent cycle. Thus, nutrition in one reproductive cycle has carry-over effects into the next.

However, for the purposes of discussion, it is convenient to consider the effect of nutrition in separate phases of the reproductive cycle on different components of sow output.

Let us therefore deal in turn with each component influencing the number of healthy weaners produced per sow per year before putting together a feeding system which will be efficient in practice.

EFFECT OF NUTRITION ON COMPONENTS OF SOW OUTPUT

The major components of sow output to be considered are:

- Number weaned per litter.
- Milk yield.
- Breeding regularity in relation to a specific age at weaning, i.e. weaning-to-conception period.

NUMBER WEANED PER LITTER

Number weaned per litter is dependent on two components, the number born and piglet mortality.

Number born

It has been shown that food or energy intake in pregnancy can vary widely (from 1·6 to 3·2 kg per day of a conventional UK diet) without affecting number born per litter. Similarly, protein intake per day in pregnancy can vary widely without affecting numbers born. Thus, it would appear that sows would have to be subjected to much lower levels of energy and protein than operate in practice before number born per litter would be reduced.

One important aspect of nutrition in relation to numbers born is 'flushing' or increasing the feed, or more specifically, the energy intake at or prior to mating. Increasing the energy intake at the time of heat in sows is unlikely to influence numbers born since, at the time of the heat, the number of eggs to be shed has already been determined. However, in the case of gilts which have been on restricted feeding, increasing the feed level by 50 to 100 per cent about ten days before service has been shown to increase number of eggs shed by about two and result in one extra piglet at birth in some trials. It would seem necessary to follow flushing with some restriction after mating.

Mortality

As has been noted in Chapter 9, pre-weaning deaths of piglets are caused by a multiplicity of social husbandry and health factors often interacting with each other. However, closely associated with mortality is birthweight as shown in Figure 11·2.

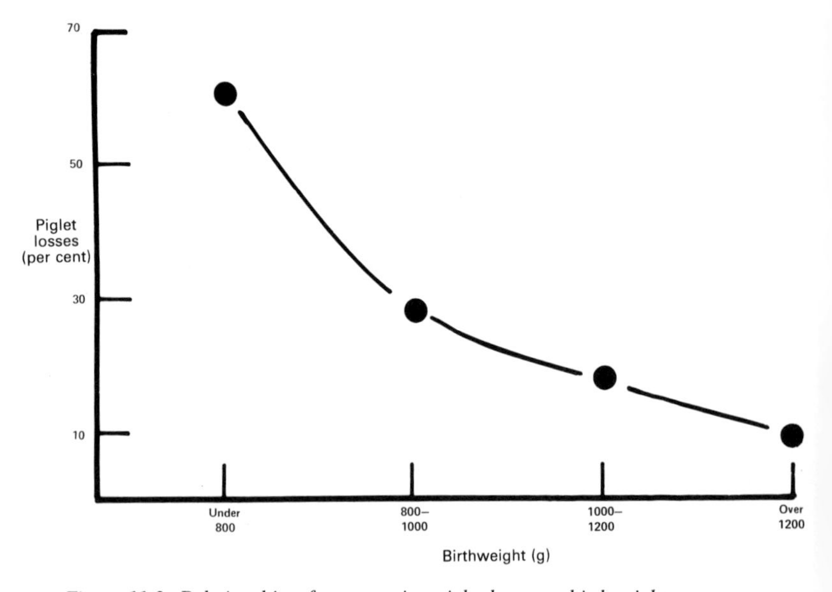

Figure 11·2. Relationship of pre-weaning piglet losses to birthweight

Low birthweight renders a piglet less viable because it loses more energy as heat to its surroundings by virtue of its proportionally larger surface area in relation to its weight. It may also have lower energy reserves at birth. Thus, the energy status of a smaller piglet is much more critical than that of a larger one. From the point of view of improving piglet survival, it is thus important to attempt to improve average piglet birthweight.

Increases in birthweight can, in fact, be achieved by increasing food or energy intake in pregnancy, but considerable increases in food intake are required to bring about very modest improvements in birthweight. It has been estimated that, to improve average piglet birthweight by only 0·1 kg, an extra 100 kg of sow food would have to be given in pregnancy to improve mean piglet birthweight and piglet survival will depend on the cost of the extra food required and the value of the extra piglets likely to be saved.

It is likely that only in herds with very low birthweights, and consequent high piglet losses, would it be worth considering increasing sow-food intake in pregnancy. There is no doubt that it would be much more cost effective to improve the farrowing facility and the environment for baby piglets and improve piglet survival in this way.

Consequently, while increasing feed intake in pregnancy can improve birthweights and therefore piglet survival, the considerable amount of extra food required to achieve a very modest increase in birthweight rarely justifies such action in the practical situation.

MILK YIELD

Piglets are entirely dependent for their nutrition in the first two or three weeks of life on the milk of their dam. Therefore, growth rate can be improved in this early period of life with increases in sow milk yield.

Milk yield can be increased both by improving nutrition in the preceding pregnancy or in lactation itself.

Higher feed levels in pregnancy resulting in increased body weight gains by the sow can improve milk yields in the following lactation and the sow can utilise these body reserves deposited during pregnancy fairly efficiently to meet the demands for milk production. However, the higher the food intake in pregnancy, the lower the appetite in the following lactation.

It is preferable, on balance, to avoid over-feeding in pregnancy so as to improve the appetite of the sow in lactation and allow her to meet demands for milk production mainly from the food consumed in lactation rather than largely from body reserves deposited in pregnancy. The aim, therefore, is to have the sow in fit but not fat condition at farrowing and to feed liberally in lactation. Overfat sows also tend to have more difficult farrowings and seem to be more prone to agalactia.

While the sow has a strong impulse to mobilise her body

reserves to meet the demands for milk production, i.e. to milk off her back, under-feeding of both energy and protein in lactation results in reductions in milk yield. A minimum daily intake of some 700 g of crude protein and 30 g lysine (provided by 5 kg of a diet with 14 per cent crude protein and 0·6 per cent lysine) is required in a six-week lactation to allow adequate milk production and this should be associated with a minimum average intake of 240 g of crude protein per day in pregnancy (e.g. provided by 2 kg of a diet with about 12 per cent of crude protein). The desirable energy or feed intakes in lactation differ from unit to unit and will vary according to the size of sow, the feeding level in the previous pregnancy, the environment and other factors to be discussed in more detail later.

BREEDING REGULARITY

In order to maximise number of litters per sow per year on a given age at weaning, the interval from weaning to conception must be minimised. Thus, the feeding system should aim at prompt appearance of heat after weaning and the maximum conception rate at this service.

Work in Ireland has demonstrated the importance of the level of protein provided in lactation for minimising the period from weaning to conception. Results of this work are presented in Table 11·2.

TABLE 11·2. Effect of protein and lysine levels in lactation on the weaning to remating interval

	Diets		
	1	*2*	*3*
Crude protein in diet (per cent)	9·3	11·8	14·3
Crude protein per day (g)	426	552	688
Lysine per day (g)	19·7	29·0	39·0
Parity No.	*Weaning to mating interval (days)*		
1	29	14	9
2	12	7	7
3	8	8	5

Source: O'Grady, J. F. and Hanrahan, T. J. (1975) Irish Agricultural Institute, Fermoy, Co. Cork.

The feeding plane during lactation in the above trial was a maximum of 5 kg per day in the first and 6 kg per day in the second and third lactations. It can be seen that the diet providing the lowest protein and lysine levels resulted in much longer weaning to remating intervals after the first lactation. This effect was much reduced following the second and third lactations.

Thus, the quality of the lactation diet in terms of protein and amino acids appears to be very critical in relation to the weaning to remating interval of the sow after weaning her first litter. This problem of rebreeding following feeding of a poor quality diet in lactation is likely to be associated, at least partly, with the greater loss of body condition of the sow during lactation when on such a poor quality diet.

Workers at Nottingham have demonstrated the importance of feeding level after weaning for achieving prompt rebreeding of the sow after weaning her first litter. In this work, gilts were fairly severely reduced in condition after weaning their first litter and they responded to increased feed levels as demonstrated in Table 11·3.

TABLE 11·3. Effect of feeding levels after weaning the first litter

	Feed levels (kg)		
	1·8	2·7	3·6
Proportion mated within 42 days of weaning*	67	75	100
Proportion farrowing	58	75	100
Interval weaning to first heat (days)	22	12	9

*Those not coming in heat within 42 days of weaning were slaughtered.

Thus, higher feed levels after weaning resulted in a higher proportion of first litter sows being mated within a given period following weaning and a shorter weaning-to-mating interval for these sows.

This effect of feed level after weaning has not been demonstrated on older sows, but it seems very desirable that every effort should be made to avoid sows being too reduced in condition at weaning and when this happens, sows should be fed liberally from weaning to service.

What must be avoided at all costs is the 'thin sow syndrome' which develops if a sow loses too much body fat. Although such emaciated sows often have big appetites, they do not readily recover condition even when on high feeding levels. Infertility or death is the more usual outcome.

A PRACTICAL FEEDING SYSTEM—THE ESSENTIALS

In evolving a sound feeding system for sows, it is logical to consider the various stages in life up to puberty and the separate phases of the reproductive cycle. The separate stages to be considered are as follows:

- Rearing the gilt.
- Feeding the gilt immediately before and after the first service.
- Pregnancy.
- Just before and after farrowing.
- Lactation.
- Weaning to remating.

REARING THE GILT

Nutrition during rearing must aim at stimulating breeding activity (puberty) in the gilt at a fairly early and predictable age and must prepare the animal for a high output in the first litter and subsequently.

One must avoid overfatness in gilts since this is not conducive to high fertility. Therefore gilts should be fed on restricted levels in the latter period of rearing. Regarding the effect of food or energy intake on age at puberty, it appears that normal levels of food restriction over the 55 to 90 kg liveweight range for bacon pigs are unlikely to delay puberty by more than one week relative to *ad lib* feeding over this period.

While much work remains to be done on the effects of specific nutrients, in particular amino acids, during rearing on age at puberty, it is unlikely that nutritional factors will have a marked influence on age at puberty, unless they are correcting an obvious deficiency.

Thus, if gilts are reared on a well-balanced diet and a feeding regime suitable for the finishing of their contemporaries destined for slaughter, it is unlikely that such a feeding regime can be greatly improved in terms of inducing earlier puberty.

FLUSHING OF THE GILT

Gilts are normally mated at either second or third heat. In preparing them for mating and pregnancy, the most effective treatment is to limit their food intake in the latter part of rearing, to increase their feeding level about ten days before mating (flushing) and to reduce feed allowance to normal restricted levels immediately after mating.

The reason for such a feeding system is that both *ad lib* feeding during the later part of rearing resulting in overfat gilts at breeding and high-level feeding after mating result in increased losses of embryos. Raising feed level by 50 or even 100 per cent from normal restricted amounts for about ten days before mating has been shown to increase ovulation rate by up to two ova and this has resulted in an increase of up to one piglet born in some trials.

This process of flushing is most effective in increasing numbers born to gilts when it follows a period of restricted feeding. The effect of flushing applied in this way varies between herds, but the response to flushing is likely to be greatest in herds where average litter size born to gilts is low, that is, in the situation where an increase is most required and welcome. A system for making it easier to implement flushing in practice was suggested in Chapter 5. It is important to revert to normal feed levels at mating as otherwise higher embryo losses can result. The investment in feeding an extra 10 to 20 kg of food in the ten days or so prior to mating the gilt is not high and can often produce a very worthwhile return.

PREGNANCY

It has already been established that while increased feed intakes in pregnancy can increase birthweight of piglets, a considerable increase in food intake is required to bring about a relatively small improvement in birthweight. Since improvements in piglet survival can be achieved more effectively by improving the farrowing environment rather than by increasing birthweight, attempts to improve birthweight and piglet survival by increasing food intake in pregnancy can rarely be cost effective.

Higher feed levels in pregnancy result in greater body weight gain of the sow. The sow can mobilise these body reserves deposited during pregnancy or 'milk off her back' to meet some of the demands for milk production. However, it is preferable to meet as much of the demand for milk production as possible by adequate feeding during lactation. A sow which is fed too liberally during pregnancy will have a lower appetite in lactation.

The pregnant sow is a more efficient converter of food than the non-pregnant animal and this should be kept in mind when deciding on the desirable level of feeding in pregnancy in any given situation. Pregnant sows should receive a minimum of 240 g of crude protein daily but one cannot be specific about energy or feed intakes since these will vary with the size of sow, the temperature level in the house and other factors to be discussed later. The general aim of pregnancy feeding should be to have sows in fit but not fat condition at farrowing.

Regarding the desirable distribution of food over pregnancy, some have suggested that higher levels are desirable in early pregnancy, lower levels in mid-pregnancy with an increase in feed intake towards the end of pregnancy. However, work at the Rowett Research Institute and other centres has demonstrated that, provided the same total amount of feed is given, it does not matter how this is distributed over the different phases of pregnancy. So, by keeping a constant level of feeding throughout, as good results are produced as by any departure from this programme and, of course, this helps to simplify feeding management.

It is obviously desirable to feed pregnant sows according to their body condition so that both under- and over-feeding are avoided and all sows attain the same target body condition at farrowing. Thus, individual feeding for most of pregnancy is very desirable.

FEEDING AROUND FARROWING

The tendency is to impose quite severe restrictions on food intake in many herds when feeding around farrowing. This practice is partly related to the fear of increasing the risk of conditions leading to agalactia or milk shortage if higher levels were given and partly to the belief that the appetite of the sow is low at this time. The fact is that some sows have substantial appetites just after farrowing and such sows may become very restless and thus endanger their newborn piglets if underfed. The relevance of underfeeding at this time and consequent restlessness of the sow in relation to increasing piglet losses was discussed in some detail in Chapter 9.

While many hold the view that excessive feeding around farrowing is likely to induce conditions leading to agalactia or milk shortage at this time, this relationship has not been confirmed experimentally.

A few producers practise low-level feeding in general around farrowing, but sows with apparently keen appetites are given considerably more food. We feel that this is the correct way to go about things as the hungrier sows will be kept more satisfied and are likely to be less restless and less likely to crush piglets as a result.

FEEDING IN LACTATION

Food requirements during lactation depend on the number of piglets being suckled as a greater number of piglets in the litter will stimulate greater milk output from the sow. Requirements will also vary according to the size of the sow and on the temperature level in the house.

Sows will, almost inevitably, milk off their backs to some extent and thus lose weight in lactation. The extent of this weight loss must be controlled by adequate feeding in lactation; otherwise there may be problems with rebreeding and even the 'thin sow syndrome' could result.

Nutrition in lactation must therefore cater adequately for the milk output of the sow. Milk output is relatively low at the start and gradually reaches a peak some three weeks after farrowing. As a result, weight loss tends to be greater after three weeks than in the first three weeks of lactation. Thus, daily nutrient requirements for sows being weaned at two to three weeks of age will be lower because of lower average daily milk production than for those weaned between four and six weeks of age or later.

A sow rearing a litter of about nine piglets up to weaning at four to six weeks should have a minimum daily intake of protein of around 700 g and minimum lysine of around 30 g. This can be provided by a diet with 14 per cent of crude protein and 0·6 per cent of lysine if the sow consumes 5 kg of this diet daily.

Inadequate amounts of protein or lysine or energy will result in greater loss of body condition in lactation. The later the weaning, the higher the desirable food intake to avoid excessive weight loss by weaning. Adequate food levels must therefore be provided in lactation to ensure that the sow is in reasonable condition at weaning.

However, offering enough food is one thing and getting the lactating sow to eat what she is offered is quite another. Increasingly, producers have problems in encouraging sows to consume enough in lactation to avoid excessive loss of body condition by weaning.

It may be that modern improved sows have lower appetites than their relatively unimproved counterparts. However, even if this is so, there are several steps one can take in an attempt to improve food intake in lactation.

FACTORS AFFECTING APPETITE IN LACTATION

Feed intake in pregnancy

As discussed previously, it is well established that the higher the food intake in pregnancy, the lower will be the appetite in lactation. Thus, overfeeding in pregnancy must be avoided.

Environmental temperature
Sows will consume more at lower room temperatures as demonstrated by work carried out in Ireland and presented in Table 11·4.

TABLE 11·4. Effect of room temperature on food intake in lactation*

	Temperature of farrowing house	
	27°C	*21°C*
Feed intake per day (kg)	4·6	5·2
Weight loss of sows (kg)	21	13·5

*Weaning took place at 31 days.

Source: Lynch, P. B. (1977), Irish Agricultural Institute, Fermoy, C. Cork.

Thus, if higher food intakes are required, it is desirable to provide creep rather than whole-house heating so that the sow is kept in cooler conditions. If space heating is used and the temperature at piglet level is around the optimum for newly-born piglets of about 28 to 30°C at farrowing, this can be reduced after the first few days so as to stimulate higher food intake of the sow.

Energy level of diet
Results in Ireland have also shown that sows will consume the same amount of a high-energy diet as of a more conventional one. Thus, total energy intake can be increased by providing a higher-energy diet, and so these diets can play a useful part in situations where low intakes in lactation are a problem.

Wet versus dry feeding
Sows will normally consume more food in wet (about 1 part food to 2½ parts water) than in dry form.

Cubes versus meal
Feed is more readily eaten in cubed than in meal form.

Frequency of feeding
Sows will consume more if fed twice- than if fed once-daily. Consumption is likely to increase further if they were fed even more frequently but this would increase labour costs. However, if lactating sows had a small hopper fitted to the front of their farrowing crate, they could be fed *ad lib*. On this system they will tend to feed little and often. *Ad lib* feeding could start from soon after farrowing or, in the case of those who find high-level feeding around farrowing to be associated with a higher incidence of

conditions leading to agalactia or milk shortage, the start of *ad lib* feeding could be delayed for a few days.

Thus, there are several ways of trying to overcome a low appetite problem in lactation and avoiding excessive loss of body condition by weaning.

FEEDING FROM WEANING TO REMATING

The main objective of feeding in this period must be to stimulate prompt breeding activity after weaning, a well-exhibited oestrus so that it can be readily detected and an adequate ovulation rate.

The importance of a diet of adequate protein quality in lactation for shortening weaning-to-mating interval, after weaning the first litter in particular, has already been discussed (see Table 11·2). The effect of increasing the feed level after weaning on shortening the weaning-to-service interval in the case of gilts which were much reduced in condition by weaning has also been covered (see Table 11·3).

There are a great variety of feeding practices at weaning carried out on commercial farms aimed at drying the sow off as quickly as possible and stimulating prompt breeding activity thereafter. Some starve the sow of both food and water on the day of weaning while some withdraw food only. Others provide normal feed levels of 2 to 2·5 kg daily from weaning, while some producers leave sows on their high lactation feed levels of 4 to 6 kg daily until they come in heat and are mated.

It is difficult to rationalise about these varied practices and the most appropriate advice would appear to be that if the particular system in operation is resulting in prompt appearance of heat in all sows after weaning, good conception rates and high subsequent litter size, then one should leave well alone.

From the physiological point of view, the most effective way to dry off the sow is to continue high-level feeding after weaning and make water freely available; and the increase in intra-mammary pressure as a result of the continued secretion of milk will very effectively and quickly stop milk secretion.

Thus, the continuation of high-level feeding after weaning should quickly dry the sow off and, if this high level of feeding is continued until mating, it is unlikely to do any harm and is likely to be beneficial for sows which have lost a lot of condition after their first lactation in particular. Since the weaning-to-remating interval should be relatively short, about five to seven days, higher intakes than normal in this period constitute a fairly small investment and a return is likely on a small proportion of sows in most herds which seem to benefit from a nutritional boost at this time.

The only other aspect of nutrition in the weaning-to-mating period worthy of mention is the possible response from extra vitamins at this stage. Some producers who include liberal quantities of white fish meal in the diet for a few days after weaning claim substantial benefits in terms of short weaning-to-conception intervals and good subsequent litter size. For the same reason, there is much support on many units also for the practice of administering a multivitamin injection to sows at weaning.

CHECKING ADEQUACY OF FEEDING PROGRAMME

So far, the main emphasis in this chapter has been in considering the effects of nutrition in separate parts of the breeding life and reproductive cycle of the sow.

However, it was stated earlier that nutrition in one phase of the reproductive cycle cannot be considered independently from that in another phase as there are carry-over effects from one phase to another and from one cycle to another.

It has also been emphasised throughout the text that it is impossible to generalise on desirable levels of nutrients and feed in general since the optimum level of feeding varies according to such factors as:

- The size of the gilt at first mating and of the sow subsequently.
- The environment provided.
- Method of feeding (individual or group).
- The health of the herd.
- The level of productivity of the sows.
- The standard of management.

Food requirements will increase with increases in size of the gilt and sow, in a poorer environment, on group feeding, in a less healthy herd, in a herd with a higher output of weaners per sow per year and under poorer management conditions.

Because these criteria governing food requirements will vary so markedly from unit to unit, it is impossible to prescribe recipes of desirable feed levels in different stages of the reproductive cycle and breeding life of a sow and for a herd in general.

In view of the difficulty of generalising on desirable levels of feeding, another method has been evolved to check the adequacy of existing feeding programmes and for providing guidelines for new ones. This is based on checking liveweight changes of sows from one reproductive cycle to the next. The system was suggested some time ago by workers then operating at the Rowett Research Institute and Nottingham University. The recommendation is that sows should gain between 10 and 15 kg from farrowing to farrowing or from weaning to weaning. This increase in liveweight should proceed up to about the fourth or fifth litter after which liveweight should stabilise. The recommendation is summarised in Figure 11·3.

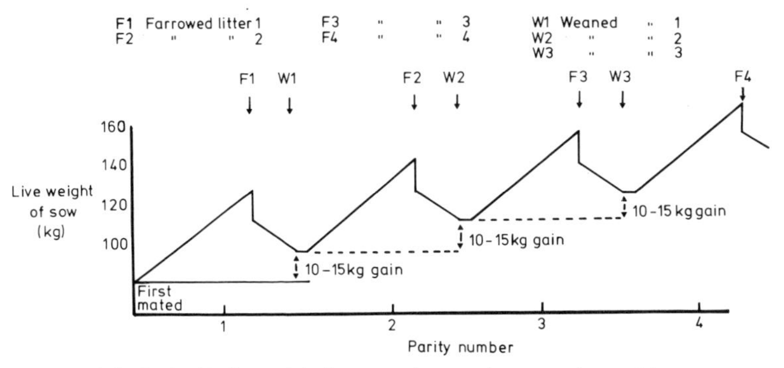

Figure 11·3. Desirable liveweight increases in sows in successive parities. (Interpolated from the work of Professor G. A. Lodge and the late Professor F. W. H. Elsley)

The normal sow is constantly changing in liveweight as she gains weight in pregnancy, takes a sudden drop for obvious reasons as she farrows, loses further weight in lactation as she mobilises her body reserves to help meet the demands for milk production, before gaining weight again as she proceeds on her next pregnancy. Checking that the sow is gaining from litter to litter by the recommended amounts is best done by weighing sows at each weaning.

The desirability for a 10 to 15 kg liveweight increase in each of the first four litters is because we are starting off with an immature animal as a gilt which continues to mature up to about the fifth litter.

During this time the skeleton is growing all the time and allowance for this level of liveweight gain will permit the same body condition to be maintained and so guard against losses in performance. If the sow is gaining less from litter to litter, she will be gradually becoming more of a skeleton—a higher proportion of bone and a lower proportion of muscle and fat. If she is gaining more than 10 to 15 kg, she is being overfed and food is being wasted for she will not be more productive as a result.

Having established the target for weight change over a complete reproductive cycle, it is then a question of how best to distribute the food required to bring about this desired weight increase. Normally, it will be best to offer very high levels of a well-balanced diet in lactation (e.g. 6 kg per day) and only very moderate levels in pregnancy (e.g. 2 kg per day).

Alternatively, if there is a low birthweight and resulting high mortality problem in the herd, then it might possibly be advisable to increase the pregnancy level somewhat and correspondingly reduce the level fed in lactation. The precise feed levels required to bring about the desired weight change from one cycle to the next will vary from farm to farm according to the criteria already discussed.

Having decided on the feed required to bring about the desired weight increases and how best to distribute this feed between pregnancy and lactation according to the particular requirements on the farm, it will then be a matter of modifying the routine to get the best out of stock. Such modifications of the normal routine have already been discussed and are summarised below; for example, in the flushing of gilts, in putting gilts and sows which are suckling large litters on to higher feed levels or on to *ad lib* feed during at least part of lactation to avoid excessive weight loss, and in feeding fairly liberal amounts from weaning to service to encourage a prompt heat after weaning.

It is a case of treating sows as individuals and feeding them so that they neither become too big, which is a waste of feed, or fall to excessively low condition which, if not resulting in the 'thin sow syndrome', is likely to give rise to rebreeding problems.

FEEDING REGIME—SUMMARY

Protein and amino acids

Taking a conventional UK diet based on barley and soya bean meal as presented in Table 11·1 (possibly with the addition of a small amount of white fish meal if the price is competitive with that of soya), this should have about 14 per cent crude protein and about 0·6 per cent lysine in lactation and this can be reduced slightly in pregnancy. However, the total amount of protein and lysine fed per day is more important than the per cent inclusion rate in the diet. For example, 6 kg daily of a diet with 14 per cent crude protein and 0·6 per cent lysine will provide 840 g protein and 36 g lysine daily, whereas 4 kg of a diet with 16 per cent crude protein and 0·7 per cent lysine will provide only 640 g and 28 g protein and lysine respectively. In lactation, minimum daily quantity of crude protein should be 700 g and of lysine 30 g (can be supplied by giving 5 kg of a diet with 14 per cent crude protein and 0·6 per cent lysine). In pregnancy, minimum daily quantities of crude protein and lysine should be 240 g and 12 g respectively.

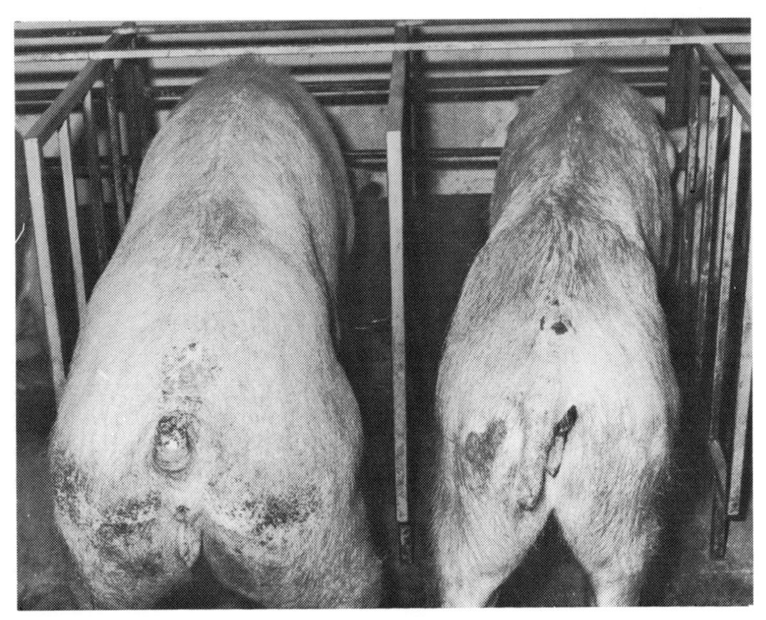

Plate No. **42**

Sows must be treated as individuals and the ration varied according to body condition.

Using a conventional UK diet, essential amino acids other than lysine will normally be supplied in adequate amounts.

Diets deficient in protein and amino acids in pregnancy and lactation will result in greater weight loss in lactation. It must also be remembered that where inadequate quantities of food are being offered, sows will use protein and amino acids as an energy source. In a very warm farrowing house, sows may have poor appetites which will reduce their intake of protein and amino acids. For these various reasons producers prefer to supply reasonable margins of safety in the diet in terms of protein and amino acids.

Energy or feed level

Optimum feed level will vary according to the adequacy of housing, the size of sow, milk output and other factors as previously discussed. The general aim should be to allow the sow to gain between 10 and 15 kg liveweight from one weaning to the next or from one farrowing to the next. This increase will help to keep the sow in the same body condition as it matures. The sow should be fully mature at about the fifth litter and no further increase in liveweight is required after this stage.

Normally, the best way to distribute feed is to feed to appetite in lactation and feed only moderate levels in pregnancy, the particular level for each farm being geared to keeping the sow in reasonable body condition and achieving the desired increase in liveweight from one litter to the next.

Vitamins and minerals

Vitamin/mineral supplement for sow diets should be included at the rate recommended by reputable compounders.

Rearing of gilts

In the later part of rearing, gilts should be on restricted feed levels to avoid over-fatness at service. Gilts from strains which have not been genetically improved in terms of higher lean tissue growth rate and reduced fat deposition may have to be restricted from around 55 kg liveweight, whereas gilts from genetically improved lines need not be restricted until a much later stage in growth.

Flushing

Normally gilts are served at either second or third heat. When gilts which have been reared on restricted feed levels have their feed level increased by 50 per cent or even doubled in the ten days or so before service, this has been shown on some units to increase numbers born by about one piglet per litter. Flushing for a short period in this way is not costly and it may produce a worthwhile return on some units.

A system for making it easier to flush gilts in practice was suggested in Chapter 5.

Feed level of flushed gilts should be reduced immediately after mating as high level feeding in early pregnancy can increase losses of embryos.

Increasing the feed levels of sows on the day of mating is unlikely to increase the numbers born.

Feeding around farrowing
Some claim that, under their conditions, high-level feeding around farrowing stimulates conditions which lead to agalactia or milk shortage. This association does not hold on other units. If feed level just after farrowing is too low in the case of sows with good appetites, this can result in increased restlessness of sows and an increase in piglet deaths from crushing. Thus, a balance has to be kept in relation to feeding level around farrowing.

Feeding in lactation
The aim should be to avoid sows being too reduced in condition at weaning as, otherwise, rebreeding problems could result. Thus, sows should be offered liberal quantities in lactation and, if necessary, *ad lib* fed by fixing a small hopper to the front of the farrowing crate. In the case of those units which experience difficulty in getting sows to eat enough in lactation, attempts can be made to increase appetite of the sow by the following methods:

- Reduce the room temperature level for the sow (maintaining warmth in the creep for the piglets).
- Feed wet rather than dry.
- Feed cubes rather than meal.
- Feed more frequently or, more simply, feed *ad lib*.
- Use a higher-energy diet.

Feeding weaning to service
The objective in this period is to minimise the period from weaning to service and conception so as to maximise the number of litters per sow per year.

Ensuring an adequate protein and lysine level in the previous lactation has been shown to be of crucial importance in minimising the period from weaning to oestrus, especially in the first lactation, as an adequate quality of diet in lactation helps to reduce body weight loss.

For gilts which have lost a lot of condition in lactation, high feed levels (around 4 kg per day) should be maintained after mating as this will help to minimise the weaning-to-mating interval. Maintaining high feed levels after weaning helps to dry the sow off quickly and this nutritional boost is likely to benefit a proportion of sows which have milked heavily in lactation. The period from weaning to mating should be short, so the extra costs of a higher level of feeding should not be high and is likely to leave a worthwhile return in the case of at least some sows in most herds.

Routine provision of a multivitamin injection to all sows at weaning is a nutritional practice from which an increasing proportion of producers claim advantage for minimising the weaning-to-conception period.

GETTING THE BEST OUT OF A SOUND FEEDING POLICY

One is not belittling the tremendous strides made by nutritionists, but rather praising these, by stating that feeding is now pretty straight-forward. The principles are readily grasped by the good stockman who adapts them to suit his own particular conditions. He will avoid over- and under-feeding, he will keep his sows in the right condition and he will feed liberally at strategic points in the cycle. Thus, he will have decided on his optimum feeding strategy to suit his conditions. This will be a useful start.

What the stockman has to do next is organise all the other vital inputs such as stock, the environment, disease prevention measures, etc. and exploit them to the full, using all his husbandry skills to ensure that adequate numbers are born, there is maximum piglet survival, prompt appearance of heat after weaning and effective service.

Only when this skilled attention is given to all the other factors will a sound feeding policy be fully exploited to maximise the margin between value of weaners produced and the food costs involved.

Chapter 12

ESTABLISHING THE NEWLY WEANED PIG

How BEST to manage the newly weaned pig is outside the scope of this book. This subject will be comprehensively covered in the book *The Growing-Finishing Pig: Improving its Efficiency* by P. R. English, S. H. Baxter, V. R. Fowler and W. J. Smith, to be published by Farming Press Ltd.*

However, management during suckling can have a major influence on the ease or difficulty of management after weaning. Consideration of how best to manage litters during the nursing stage in relation to minimising problems and easing management at weaning forms the basis of the present chapter. Major consideration will be given to feeding in relation to minimising nutritional changes at weaning and achieving a gradual safe transition from the pre-weaning to the post-weaning diet.

NUTRITIONAL CHANGES IN RELATION TO AGE AT WEANING

Until recent times, the standard age at weaning was eight weeks. This made sense from the point of view of minimising the change in diet at weaning because in the final week prior to weaning the piglet would be obtaining 70 to 80 per cent of its food requirements from creep food, the remainder coming from milk. Consequently, when the same creep food continued to be fed after weaning, the piglet was not subjected to too great a change in diet. When weaning is at six weeks, piglets in the final week prior to weaning will normally be obtaining between 50 and 60 per cent of their food requirements from creep food and the remainder from milk. Thus, weaning at six weeks entails a greater dietary change for the piglet and as weaning becomes progressively earlier, the dietary change at weaning is likely to be greater. If we go to the extreme of weaning as early as ten days of age, a complete change in diet is likely to be involved as the piglet will normally have consumed virtually no solid food before this stage. So the earlier one

* Scheduled for publication in 1983.

attempts to wean, the greater the efforts which should be made to encourage consumption of solid food prior to weaning so as to minimise the extent of dietary changes.

And in setting out to increase creep-food intake prior to weaning, one must always bear in mind that creep-food consumption varies markedly both between litters and between piglets within litters so that some piglets will have consumed very little solid food prior to weaning, whatever the age at which this takes place.

STRESS AT WEANING

Even without dietary changes, weaning involves stress, the extent varying according to the age at weaning and the management care exercised. Stress is known to have a general adverse effect on all body functions, including the digestive system. The natural contracting movements of the stomach slow down and complete cessation of stomach movement (stasis) is common. There is an increase in blood flow to the gut as a whole, leading to congestion of the blood vessels supplying it, and the lining itself may be involved in small haemorrhages and ulceration. Stress also affects the production of certain hormones and this reduces the resistance of the animal to disease. (For further reading on stress, see Chapter 3.)

UPSET AT WEANING ASSOCIATED WITH DIETARY CHANGES

The stress of weaning is further accentuated if piglets are subjected to marked changes in diet at this stage. A change in diet at weaning is likely to result in the following digestive upsets in normal animals to a greater or lesser degree:

- An increase in excretion of fatty acids in the faeces.
- An increase in the output of carbohydrates in the faeces.
- More watery faeces.
- In some cases there is an increase in the multiplication of *E. coli,* especially the haemolytic strains.
- Degenerative changes in the cells lining the gut.

These changes occur shortly after weaning and reach a peak in seven to ten days. So after weaning and a change of diet, each piglet becomes a potential scourer. All this adds up to what we can call temporary indigestion or, in more exact terms, 'malabsorption syndrome'. Should scouring begin, this damages the gut lining and is liable to upset the production of IgA (immune globulin A) from the gut wall. The diminished production of IgA, a substance which normally protects the gut lining, places the animal at further risk. Certain potentially pathogenic *E. coli* may take advantage of this situation and multiply rapidly. They produce toxins which not only cause further degenerative changes of the gut, but are also absorbed and may cause damage to other organs in the body, e.g., the brain in bowel oedema. This is the reason why it is so important to minimise dietary change at weaning and to minimise stress by good management at this time.

IMMUNITY IN RELATION TO AGE AT WEANING

The piglet weaned around three weeks of age or before has still further problems with which to contend and these refer to its immunity or resistance to disease at this time.

The piglet is born without any protective immunity and this is obtained from the special proteins (gamma globulins) from the colostrum or first milk of the sow. This circulating immunity protects the piglet for its first ten to fourteen days of life, at which time it declines to negligible levels (see Figure 12·1). The piglet

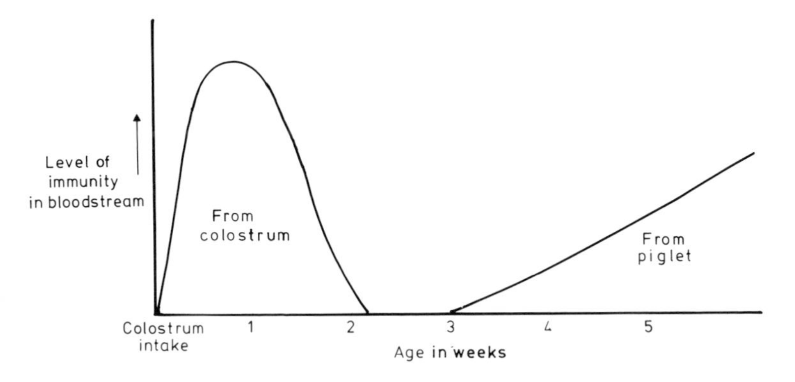

Figure 12·1. Type and duration of protective immunity.

does not normally start to build up its own immunity to prevailing infection until about three weeks of age and the level of immunity builds up rather slowly from this point.

Thus, it can be seen that early weaning (ten days to three weeks) results in the piglet being weaned at a stage when its immunity is at the lowest level in its life (other than at birth). The piglet which is left on the sow has an added advantage because special 'immune protein' (immune globulin A or IgA) which is constantly secreted in the milk of the sow is neither absorbed through the gut nor digested by the gut enzymes and this bathes and protects the cells lining the gut as long as the piglet is suckling. This immunity is important with regard to certain diseases such as *E. coli* disorders. While the gut wall of the piglet also secretes IgA, this source of protection can also be upset by the stress and subsequent problems associated with earlier weaning. Therefore, it follows that the early-weaned piglets will be susceptible to disease, and hence they should be subjected to special management care at this time in order to cushion them from infection.

Vaccination of the early weaner by injection is of little benefit in controlling disease because (1) the piglet is incapable of producing enough antibodies until three weeks of age and (2) two injections are necessary at an interval of ten days in order to produce a practical level of immunity.

Thus, the piglet at weaning is subjected to stress and dietary changes and, the earlier weaning takes place, the greater the challenge to the piglet. Every effort must be made at weaning, therefore, to provide the piglets with as good conditions as possible in order to reduce the challenge to the newly-weaned pig.

Provision of such conditions at weaning in terms of environment, comfort, nutrition and minimum challenge from infection will be adequately covered in *The Growing-Finishing Pig* (Farming Press).

Of particular relevance here, however, is the feeding of the piglet before weaning in relation to minimising dietary changes at weaning so as to reduce the risk of digestive upsets and associated problems at that stage.

MINIMISING DIETARY CHANGES AT WEANING

It was established earlier that the later weaning takes place, the greater the intake of solid food prior to weaning and therefore the easier it is for the pig to adapt from pre- to post-weaning nutrition. It is also clear that the higher the consumption of solid food one can encourage before weaning, whether at ten days or eight weeks of age, the more readily the pig will adapt and stabilise its digestion following weaning.

Thus, particularly for earlier weaning, the objective is to increase solid food intake prior to weaning. Difficulty in encouraging piglets to eat plenty of creep food prior to weaning is almost a universal problem among pig-keepers.

INCREASING CREEP-FEED INTAKE

In order to appreciate the secrets of achieving higher creep-feed intakes in piglets, one must have a background knowledge of the digestive system of the baby piglet and how this develops.

At birth, the piglet is equipped with the necessary enzymes to digest the nutrients in milk, i.e. casein (milk protein), lactose (milk sugar) and milk fat. The piglet can also utilise glucose at this stage and also other unsaturated fats such as maize oil and peanut oil.

Animal fats are less suitable but if such are to be used, lard is preferable to tallow.

The digestion of casein is aided by its being formed into a clot in the stomach. Other sources of protein do not clot in this way but the piglet's ability to deal with a wider range of proteins develops in the first few weeks of life. The process of hydrolysis of protein increases their digestibility. Hydrolysed meat and fish meals and solubilised soya bean protein are useful ingredients for baby pig diets.

Similarly, apart from lactose and glucose, the piglet cannot digest more complex sugars such as sugar (sucrose) and starches at first, but the enzymes necessary to digest these carbohydrates develop in the first few weeks of life. The digestibility of starch can be improved by processes such as cooking, which rupture the starch granules and allow them to be more easily attacked by the relevant enzymes. So cooked cereals such as rolled oat groats and flaked maize are useful energy sources for baby pig diets.

This natural process of development of the digestive system and of the relevant enzymes is speeded up by encouraging piglets to eat small quantities of these non-milk proteins and carbohydrates.

To encourage early creep-feed consumption, certain rules must be followed:

- It is safer to formulate diets for very young pigs using milk-based nutrients and unsaturated fats.
- Small quantities of good quality non-milk proteins and carbohydrates such as cereals (preferably pre-cooked), fish meal and soya bean meal can be offered and this will speed up

development of the digestive system and equip it to deal with a wider range of nutrients and foodstuffs.

- The diet should be fed on a 'little and often' basis so that it is always fresh and the curiosity of the piglets is being constantly stimulated by arrival of fresh material in the creep feed trough. This should have the effect of encouraging consumption.
- To encourage intake, the creep food must be attractive to the piglet and acceptability is likely to be improved by the creep food being sweet, of good texture and not too finely ground and dusty and by being always offered in the fresh state. Skim milk, with or without added fat, and also rolled oats, tend to improve the appeal of a diet to piglets and help to encourage higher intake.
- Piglets may be more attracted to diets in meal than in pellet form at first. However, it is important to avoid both a dusty meal and a hard pellet. If the meal contains such materials as rolled oats and flaked maize, this will help to give it the open texture desirable to piglets. Alternatively, if pellets are used, a small quantity of skim milk scattered over the top of the pellets can be very effective in attracting piglets and encouraging intake.

The optimum type of creep feed to offer will depend on such considerations as the cost of the diet or diets, the general aims in terms of target weaning weight, and, in particular, on the age at weaning.

Weaning around three weeks of age

Maximum consumption of creep food should be encouraged so as to minimise the dietary change at weaning. The creep diet to be used should be that which is to be fed after weaning. To be suitable for pigs weaned at three weeks of age, this diet should be highly digestible and should contain 20 to 30 per cent of dried skim milk, easily digested fats, cooked cereals such as rolled oats or rolled wheat and other good-quality cereals and protein sources, as well as minerals and vitamins. The diet should contain 21 to 23 per cent crude protein, 1·1 to 1·2 per cent lysine and about 16 MJ of digestible energy per kg on an 'as fed' basis. Creep feeding should start when piglets are about seven days of age.

This diet should continue to be fed for one to two weeks after weaning and, if high consumption is achieved by following the above recommendations prior to weaning, this will help to minimise the extent of dietary change and the resulting upset at and shortly after weaning.

Such pre-weaning diets will be costly but they can often be

justified (particularly where piglets are somewhat short of milk), on account of the high degree of efficiency with which they are converted and by minimising the incidence and severity of post weaning problems.

WEANING FROM FOUR TO SIX WEEKS

Farmers often experience difficulty in encouraging piglets to eat enough creep feed even when weaning as late as six weeks of age.

A suitable post-weaning diet for this age of weaner might contain around 20 per cent crude protein, 1 per cent lysine and 13·0 MJ of digestible energy per kg. It should have 5–10 per cent of dried skim milk, along with other suitable ingredients such as fish protein, soya bean meal, microbial protein, rolled oats, barley, flaked maize, sucrose (not more than 10 per cent), minerals and vitamins.

This diet should also have been provided previously as a creep food so as to minimise the dietary change at weaning. Whether a largely milk-based starter diet is offered from an early age in this case depends on its cost in relation to its estimated benefit (see Chapter 9 page 215).

If such a milk-based diet is relatively cheap in relation to its beneficial value and/or an increase in creep consumption and weaning weight is desirable from the point of view of easing the transition and minimising problems at weaning, then the use of such a milk-based starter diet is likely to be justified.

WEANING FROM SIX TO EIGHT WEEKS

Usually creep-feed consumption prior to weaning is fairly considerable in litters weaned at this late stage and so severe dietary changes at weaning are avoided and the risk of post-weaning problems much reduced.

A suitable post-weaning diet in this case would contain 18 to 20 per cent crude protein and 0·9 to 1·0 per cent lysine. It could contain a token quantity of skim milk (say, 5 per cent) and otherwise consist of such ingredients as ground barley and maize, rolled oats, good-quality proteins such as white fish meal, and soya bean meal as well as vitamins and minerals. The same diet would be suitable for some time prior to weaning and whether or not a largely milk-based starter creep be fed at an earlier stage is dependent on the same considerations as outlined above for four- to six-week weaning. Feeding such a starter creep is, however, slightly less justifiable for six- to eight-week weaning because creep food intake is likely to be sufficiently high at weaning at this stage to minimise the risk of digestive upsets subsequently. In addition, offering improved creep foods, although it may increase weaning weight, may not result in this advantage being maintained up to the slaughter stage.

Figure 12·2 summarises in diagrammatic form the foregoing comments and represents possible creep-feeding schemes designed to achieve reasonable creep-feed intake and reduce dietary changes and digestive upsets at weaning.

'STARTER' AND 'FOLLOW-ON' CREEP DIETS—GENERAL CONSIDERATIONS

Experience in the field varies widely regarding the general usefulness of using two creep diets, i.e. a high-quality 'starter' and

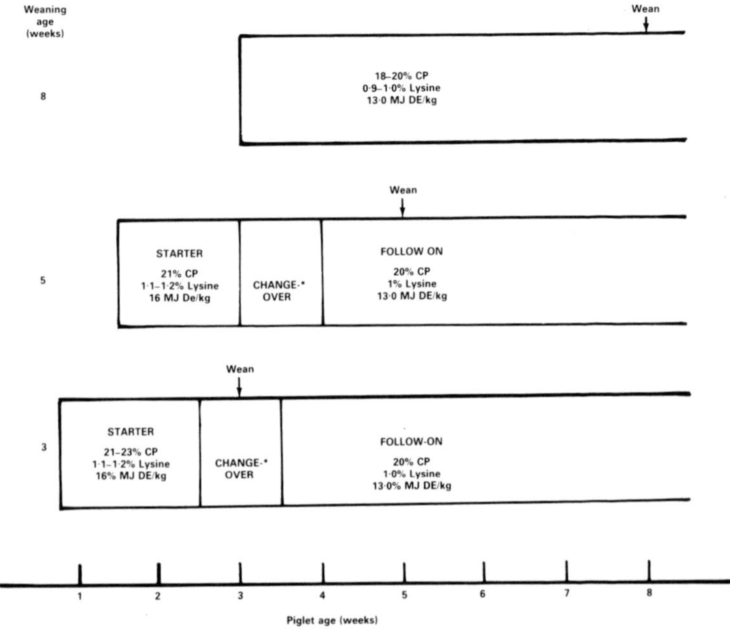

Figure 12·2. Practical creep feeding schemes designed to achieve high intake of solid food prior to weaning and to minimise dietary changes at that stage.
**Change-over period involves adding increasing proportions of 'follow-on' to 'starter' feed.*

a lower-quality 'follow-on'. Mainly for convenience, some use a 'follow-on' creep diet only. Others acknowledge that a 'starter' diet has some initial merit, but, if weaning at five weeks or later, find that piglets receiving only 'follow-on' creep tend to compensate later for slower earlier growth. Other producers, particularly those practising earlier weaning, find the use of a 'starter' diet to their advantage.

An alternative practice to employing two 'creep' diets which some producers find useful is to use a 'follow-on' diet in conjunction with piglet milk substitute or skim milk powder. The powder can be mixed in fairly high proportions with the 'follow-on' creep at first and its concentration gradually reduced as the piglets begin to consume increasing quantities of creep. Higher proportions of such milk powder can be maintained in the creep feed for litters on poorer-milking sows and in the case of litters containing one or two poorer-thriving piglets. Such milk powder provided on a 'little and often' basis is very attractive to young piglets and helps to encourage earlier intake of creep food.

WATER INTAKE

It is important that fresh water is made readily available to young piglets if high intakes of creep food are the general aim. Provision of a nipple drinker in the farrowing pen will help to ensure that piglets can use such, if this type of drinker is the one used after weaning. Inadequate water intake can be a problem after weaning and training to drink is as important as training to eat solid food in relation to reducing post-weaning problems.

RESTRICTION OF FOOD INTAKE AFTER WEANING

If piglets have consumed very little solid food before weaning, the extent of dietary change at weaning can be quite severe. If, however, the post-weaning diet is entirely milk based, then the transition from almost an entirely sow milk diet will be easier. However, if the post-weaning diet is offered *ad lib,* some pigs may gorge themselves. Such over-indulgence can lead to digestive problems and scour caused by an overloading of an immature digestive system. Food is digested in the stomach and small intestine and, if food intake is excessive, undigested food can spill over from the stomach into the small intestine and from there into the large intestine (hind gut). Undigested food in the large

Plate No. **43A**

A recently weaned pen of pigs with two (arrowed) with ears beginning to flop and show the wasting or dehydration signs characteristic of post-weaning diarrhoea.

intestine can form a useful food source for bacteria normally resident there. The sudden multiplication of these previously harmless bacteria can cause scour. Such scour can result in the salts of sodium and potassium which are secreted into the small intestine via the bile, being flushed out in the scour rather than being reabsorbed in the large intestine. This lowers blood content of sodium and potassium and tends to decrease thirst and so discourages water intake. Dehydration follows, causing the wasting and 'sunken eyes' characteristic of many cases of post-weaning diarrhoea. Thus, where sudden changes of diet take place at weaning, the result can be over-consumption of food in relation to the capacity of the digestive system to deal with it, and post-weaning diarrhoea, and often death, is the final consequence.

The problem is further accentuated if unsuitable diets are fed to early-weaned piglets. For example, the digestive system of piglets weaned at ten days is equipped to deal adequately only with milk protein, lactose and fat. If such materials as starch and sucrose (sugar) are included in diets for such early-weaned pigs, these

Plate No. **43B**

Another recently weaned piglet with typical signs of post-weaning diarrhoea.

materials cannot be digested by such piglets, they will spill into the hind gut and can cause problems leading to post-weaning diarrhoea as described above.

Even when only ingredients which the piglet can digest adequately are included in diets for early-weaned piglets and there is only minimal dietary change at weaning, digestive problems, e.g. the malabsorption syndrome already mentioned, can lead to post-weaning diarrhoea.

However, the incidence of post-weaning digestive upsets can be minimised by stimulating high consumption of creep food before weaning and continuing to feed this same food for some time after weaning. The ability of the digestive system to cope with high intakes of food after weaning is best assessed by the intake of this same food prior to weaning. Thus, when high food intakes are achieved before weaning, there should be less need to restrict intake after weaning because the digestive system will be better able to deal adequately with the food consumed after weaning. However, even although the need to restrict intake after weaning is less in this situation, such restriction may still be necessary because of the possibility of the malabsorption syndrome occurring. Consequently few producers practising early weaning risk *ad lib* feeding from the weaning stage but prefer to practise a degree of food restriction for one or two weeks afterwards. There are a few notable exceptions who pay a lot of attention to creep feeding from four to five days of age, who achieve reasonable intakes of creep feed by weaning at three weeks and who continue to feed the same high-quality feed on an *ad lib* basis after weaning. The vital reasons for their success are thought to be (1) the use of a diet suited to the specific needs of the baby pig, (2) excellent creep-feeding management which helps to achieve high intake before weaning and (3) *ad lib* feeding after weaning on the basis of replenishing hoppers daily when the previous feed is all but finished. This ensures that the food is always perfectly fresh and is quite different from filling hoppers with several days' supply of food.

In relation to level of feeding offered after weaning, it should be remembered that, the higher the intake, the greater the amount of heat produced incidentally by the piglet. This 'spare' heat production helps to lower the environmental temperature requirement of the pig and should help to reduce heating costs.

CRITERIA FOR DELAYING WEANING

Pigs best equipped to perform well after weaning are those which are consuming large quantities of solid food before weaning. There is much variation in creep-food intake between litters and, when possible, weaning of litters which have consumed little creep food should be delayed. (While this is a generally applicable rule, the reverse situation can sometimes occur, i.e., some litters which, for no apparent reason, are not doing too well on the sow may benefit from being weaned). There is also much variation in creep-food intake within litters and, while it is difficult to assess intake of individual pigs, the best available indication of creep-feed intake of individual piglets within litters is likely to be liveweight. On this basis, no piglets below a certain weight, e.g., 4–5 kg liveweight should be early weaned at three weeks of age. In order to accommodate the light piglets from certain litters when their dams are weaned, these light piglets can either be gathered together and fostered on to a newly-weaned milky sow or else they can be placed on an unweaned sow from which an equivalent number of larger piglets have been weaned.

Those practising early weaning should always be prepared to pinpoint promptly early-weaned piglets which are obviously not going to adapt to the system and return these to a nursing sow.

CONCLUSIONS

The stress on piglets associated with the act of weaning and with dietary changes can lead to severe post-weaning problems for the piglet. These problems can be minimised by ensuring that the composition of post-weaning diets is adequate for the digestive capacity of the age of the piglet involved and by encouraging piglets to consume the maximum amount of the same diet prior to weaning. Where the opportunity exists, weaning of litters of piglets and of individuals within litters which are consuming little solid food should be delayed in order to reduce the risk of digestive upsets following weaning.

The pre- and post-weaning phases in the life of the piglet are usually looked upon as being separate, which indeed they are from the physical standpoint, since piglets are usually removed to another type of house and environment and may be subjected to a marked change in diet at this time. However, it is more useful to think of these two periods as one continuous phase in terms of piglet growth, and, as such, it is vital that any changes in diet are made as gradually as possible.

Chapter 13

AGE AT WEANING—CONSIDERATIONS

PIGLETS ARE weaned commercially at any stage from a few days of age up to eight weeks and beyond. While the systems which involve very early weaning (less than ten days) have become less popular in recent years, a very wide range in age at weaning still persists in practice. In view of this very variable age at weaning, there is much debate regarding what constitutes the optimum.

THEORETICAL STANDPOINT

One of the most useful measures of productivity in the breeding herd is the number of weaners produced per sow per year and this has a major influence on profitability. Earlier weaning, in theory, is an effective way of increasing sow output as indicated in Table 13·1.

TABLE 13·1. Theoretical output per sow per year at different weaning ages

	Weaning age (weeks)				
	1	*3*	*5*	*7*	*8*
Gestation period (days)	114	114	114	114	114
Lactation (days)	7	21	35	49	56
Weaning to conception (days)	10	10	10	10	10
Total cycle length (days)	131	145	159	173	180
Litters per sow per year	2·8	2·5	2·3	2·1	2·0
Weaners per sow per year (at 9 per litter)	25·2	22·5	20·7	18·9	18·0

Thus, in theory, by reducing age at weaning by one week, an extra 0·1 of a litter or about one extra pig is obtainable per sow per year. Such a potential increase is certainly very attractive, but reductions in age at weaning must be very carefully considered as many factors have a bearing on such a proposed change.

FACTORS INFLUENCING OPTIMUM AGE AT WEANING

Among the factors to be considered in deciding on the most desirable age at weaning in a given situation are the following:

Trends in sow milk yield
As outlined in Chapter 8, sow milk yield gradually rises to reach a peak about three weeks after farrowing and thereafter declines steadily to reach a low level by eight weeks of age. This creates an argument for weaning earlier than eight weeks since the sow is producing so little milk by this stage in any case.

Reducing piglet mortality
Some put forward a case for earlier weaning in order to reduce piglet mortality. However, as outlined in Chapter 9, the great majority of losses take place in the first few days of life and so weaning as early as seven to ten days of age is unlikely to have any appreciable effect on reducing piglet losses. To have any impact on piglet losses, weaning would have to take place at birth or a few hours afterwards and such very early weaning is not yet feasible commercially.

Health considerations
As outlined in the previous chapter, the piglet suffers from an 'immunity gap' around two to three weeks of age during which it is more liable to succumb to prevailing infections. Since weaning imposes stress on the piglet (the degree varying according to the husbandry care exercised), piglets weaned around two to three weeks of age are particularly liable to succumb to various ailments.

On the other hand, medicating for various ailments may be more effective following weaning. For instance, diseases such as atrophic rhinitis can be treated fairly effectively by appropriate medication through the feed or drinking water. Since piglets may eat or drink very little prior to weaning and increase intake quickly after weaning, it follows that earlier weaning makes it easier to treat conditions such as atrophic rhinitis.

Weaning-to-oestrus interval and conception rate
If weaning takes place before about ten to fifteen days of age, there is a longer period between weaning and oestrus and poorer conception rate relative to later weaning. When weaning takes place before this stage, there also tends to be a greater variation between sows in the weaning-to-oestrus interval. When weaning takes place from about sixteen days onwards, a fairly constant and

short weaning-to-conception interval can be maintained under good management. The figures in Table 13·2 below have been extracted from the records of a large commercial unit in which sows have been weaned over a fairly wide time range.

TABLE 13·2. Weaning-to-conception interval according to stage of weaning

	Age at weaning (days)						
	10–15	*16–20*	*21–25*	*26–30*	*31–35*	*Over 35*	*Over-all*
Weaning to conception (days)	12·8 (29)*	9·9 (193)	9·4 (189)	9·6 (184)	10·0 (191)	9·2 (43)	9·8 (829)

* Figures in brackets apply to the number of litters involved.

Thus, apart from weaning as early as ten to fifteen days, there was no real difference in weaning to conception interval at subsequent stages of weaning.

Subsequent litter size in relation to age at weaning

Work at the University of Nottingham has demonstrated that, while ovulation rate is not affected by earlier weaning, there is an increase in embryonic mortality (see Table 13·3).

TABLE 13·3. Ovulation rate and embryo losses according to age at weaning

	Age at weaning (days)		
	7	*21*	*42*
Number ova shed	15·6	16·8	16·9
Number embryos 20 days after mating	9·2	11·5	13·4
Number embryos missing	6·4	5·3	3·5

Source: Varley, M. A. and Cole, D. J. A. (1975), University of Nottingham

The reason for the greater loss of embryos following earlier weaning is thought to be due to the fact that the uterus or womb takes some time to recover fully from the previous gestation, and the earlier mating takes place following weaning, the less well prepared is the uterus to accept and nourish new embryos. It appears that the uterus is not fully regenerated to fully normal function until 21–28 days following farrowing so that mating before this stage is likely to result in greater embryo losses.

The greater loss of embryos following early weaning and subsequent mating in reflected in lower litter size (see Table 13·4).

TABLE 13·4. Effect of age at weaning on subsequent litter size

	Age at weaning (days)									
	6–10	*11–15*	*16–20*	*21–25*	*26–30*	*31–35*	*36–40*	*41–45*	*46–50*	*51–55*
Number born (Source 1)	9·1	9·5	9·8	10·0	10·2	10·3	10·4	10·5	10·6	10·7
Number born (Source 2)			10·1 (167)*	10·5 (182)	10·6 (187)	11·2 (207)	11·2 (51)			

* Figures in brackets are the number of litters involved.

Source 1: te Brake (1976). Source 2: Large commercial unit.

The figures in the top line are a summary of findings from many experiments and surveys.

It is clear that earlier weaning is associated with smaller litter size as a result of increased embryo losses. However, this is not to say that herds practising earlier weaning cannot achieve high numbers born. Such large litters can be, and are, achieved following early weaning, but this requires high standards of management and the same standards of management applied to later weaning would be likely to result in even larger litters.

Housing costs

The earlier the weaning takes place, the smaller the amount of farrowing space and the greater the amount of weaner accommodation required. Thus, fewer farrowing places are required for earlier weaning but such savings have to be offset against the cost of providing the more specialised weaner accommodation required. The very young weaner must be provided with a uniform high temperature for optimum performance, while ventilation and, to a slightly lesser extent, humidity, must be under control. So, if an existing herd is considering moving to earlier weaning, this means that more sows can be put through the existing farrowing accommodation and if the cost of providing the piglet accommodation is not too expensive, then the attraction of earlier weaning increases. The higher the cost of providing specialised piglet accommodation and of running costs, the less attractive earlier weaning becomes.

Food costs

By early weaning, the sow is removed from the responsibility of providing the piglet with nutrition from the time of weaning. Thus, savings can be achieved in sow food costs as a result of early weaning. On the other hand, the early-weaned piglet must be provided with a substitute for its mother's milk. Suitable feeds for early-weaned piglets tend to be very expensive but they are utilised very efficiently by the piglet.

Whether the food cost of producing a weaner at the point of sale is increased or decreased by early weaning will depend on the balance between the value of the sow food saved and the cost of the extra 'milk substitute' required by the early-weaned pig.

Also affecting this calculation of the food cost per weaner is the number of weaners produced per sow per year.

If piglet food is very expensive relative to sow food and no great increase in sow output is likely to be achieved by weaning earlier, then there is less of an argument for early weaning. On the other hand, if the cost of weaner food is not too expensive relative to sow food and a marked increase in the number of weaners per sow per year is likely to be achieved following earlier weaning, then such earlier weaning is much more attractive.

Attitude of labour and level of husbandry skill available
Earlier weaning, to be successful, demands a higher degree of stockmanship and attention to detail for success than conventional weaning at six to eight weeks of age. If the stock person responsible is fairly resistant to change, then this is a very important factor to be considered. Thus, the attitude and skill of the available labour should be assessed carefully before taking decisions on the age at which weaning is to take place.

EARLIER WEANING–ADVANTAGES AND DISADVANTAGES

From the foregoing comments, it is clear that the problem of deciding on the optimum age at weaning for a given situation is a complex one. Obviously earlier weaning offers potential advantages in terms of more litters per year and, although there are problems in getting sows rebred if weaned as early as seven to fourteen days, providing all aspects of management are right, weaning from about fifteen days onwards does not involve problems with rebreeding and conception within about ten days of weaning. However, there tends to be a reduction in litter size the earlier one tries to wean and this reduces the potential advantage in weaners per sow per year obtained from earlier weaning.

On the piglet side, much better health control is necessary and also a much better environment, especially in terms of temperature, humidity and ventilation, is required for the pig weaned around three weeks of age than for that weaned at six to seven weeks. As a result, substantial capital investment is often required to modify existing buildings, or preferably, construct new ones, specially designed for younger weaners. There is no doubt, also, that the early-weaned piglet demands a much higher standard of management and stockmanship than the six to seven week weaner.

The process of deciding on the desirable stage for weaning will be greatly dependent on whether the unit being considered is an already established one or whether a new unit is being planned.

AN ESTABLISHED UNIT

If a unit has a fair amount of life left in its buildings and it cannot be readily and cheaply adapted for earlier weaning, it would be much better to see the present position improved and consolidated before going for earlier weaning. We know that there is a great amount of slack to be taken up on existing five to eight week weaning systems. This slack probably partly arises from a combination of a sub-optimal system and sub-optimal stockmanship. If the labour is not capable of making a job of five to eight week weaning, then going for earlier weaning could be suicidal. It should be a case of improving the present system by ensuring adequate numbers born, reducing piglet losses and minimising weaning-to-conception interval by paying all the necessary attention to detail as discussed in earlier chapters. Maybe complementary cage rearing of problem piglets from ten to fifteen days of age will—as discussed in Chapter 9—be found to be a useful way of improving efficiency and this will give the stockman experience of earlier weaning on a small scale.

When a unit approaches 100 per cent efficiency on an existing system, more consideration can be given to earlier weaning, but there are units which are producing twenty-two weaners per sow per year mainly in converted buildings on five to seven week weaning. These are obviously operating very near 100 per cent efficiency for that age of weaning. They have almost reached the ceiling and when the possibility of earlier weaning is mentioned the operators seem very unwilling to consider it, which is understandable: they have an easy-care system and in normal times they obtain a very worthwhile return.

SETTING UP A NEW UNIT

For somebody who is setting up a new unit, the situation is different and early weaning becomes much more attractive. A producer considering a new unit should try to visit existing efficient units achieving around twenty-five weaners per sow per year. If they think that they have the necessary dedication, skill and stockmanship to equal that performance and the resulting budgets show that the required return on capital will be forthcoming, then they should forge ahead.

IS THERE ANY OPTIMUM AGE AT WEANING?

This is a very difficult question to answer even if one was starting from scratch in planning a new unit and adequately skilled and motivated labour was available. The decision taken is likely to be very dependent on the relative costs of various inputs and the value of output, such as sow and piglet food, fuel and housing costs, labour costs and value of weaners prevailing at that time. If these price relationships were to change it might alter somewhat the validity of the first decision.

However, there are certain criteria which favour selecting a weaning age of between three and four weeks in preference to alternative ages in setting out to plan a new unit.

The suggestion that the three to four week stage is the preferable age at which to wean if one has decided on early weaning is based on several considerations.

WEANING AT THREE TO FOUR WEEKS: REASONS FOR PREFERENCE

The immunity gap exists around two to three weeks of age. The piglets' own immunity is beginning to increase by three to four weeks and so the piglet at this stage is likely to suffer less than at an earlier stage from the effects of the stress usually associated with weaning.

By delaying weaning until three to four weeks of age there will have been a greater opportunity to encourage higher intake of solid food prior to weaning and so reduce the extent of diet change and associated check at weaning.

Following three to four week weaning, there should be no delay in appearance of heat and the depression in conception rate which is associated with very early weaning should be avoided. In addition, numbers born are likely to be higher than following earlier weaning.

By three to four weeks of age, the piglet, as well as having better immunity and having a more mature digestive system to minimise both the extent and the effect of post-weaning stress, is also not so demanding in terms of environmental temperature. It should be possible to arrange suitable housing/environment for such a piglet by harnessing its own heat production using a kennel-type arrangement either within pens or on the verandah principle so as to minimise dependence on scarce and expensive energy for providing heat. Energy costs are certainly increasing rapidly and a two to three week weaning system is more vulnerable than three to four week weaning because of its greater dependence on expensive sources of energy to provide heat.

It is desirable to keep expensive accommodation filled to capacity and since upsets are likely to occur in the regular flow of sows coming up to farrow, such upsets are likely to result in excess sows demanding farrowing space at certain times. This accommodation problem can be alleviated by weaning some litters earlier. If two to three week weaning was the standard practice, having to wean some litters at less than two weeks of age is likely to have adverse effects on both piglets and the rebreeding of the sow. Whereas the change from a standard three to four week weaning system to weaning at around three weeks is likely to give rise to fewer problems.

The stronger three- to four-week-old pig, possessing better immunity, a more mature digestive system, and greater ability to withstand colder conditions is likely to provide the stockman with fewer problems to solve after weaning and therefore will make the job more acceptable and manageable.

Weaning at three to four weeks of age has been shown in several, although not all, studies to be the optimum age at weaning in relation to achieving the maximum number of weaners per sow per year. Although the number of litters produced are less than for two to three week weaning, this is often more than compensated for by the increased litter size and lower mortality after weaning.

SURVEY FINDINGS

The only information available on the production and profitability associated with different ages at weaning stems from findings from a variety of surveys. Findings from the Cambridge University Pig

Management Scheme survey in the period 1980–81 are presented in Table 13·5.

When the economic parameters of 'Margin per weaner', 'Margin per sow in herd' and 'Return on capital' are considered, very early weaning (less than twenty days) is shown up in a good light. Later weaning appears to be associated with progressively poorer returns. Thus, the results of this survey appear at first sight to contradict our stated preference for three- to four-week weaning rather than weaning earlier than this.

TABLE 13·5. Output and profitability of herds according to age at weaning (Results for 1980–81)

	Age at weaning (days)			
	Less than 20 (17)	*20 to 29 (23)*	*30 to 39 (35)*	*40 and over (44)*
Number of herds	17	55	33	15
Litters per sow in herd	2·31	2·24	2·05	1·94
Live pigs born per litter	10·2	10·3	10·7	10·2
Weaners per litter	8·9	8·8	8·9	8·8
Weaners per sow in herd	20·6	19·7	18·2	17·1
Weaner weights at 8 weeks (kg)	18·1	17·8	18·5	17·8
Feed per weaner to 8 weeks (kg)	73·5	79·1	88·5	88·9
Cost of feed per tonne (£)	157·76	148·48	142·54	135·30
Cost per weaner at 8 weeks:	£	£	£	£
Feed	11·60	11·74	12·62	12·03
Labour	3·15	3·54	3·48	4·46
Other costs	3·97	3·92	3·87	3·87
Stock depreciation	0·50	0·36	0·33	0·24
Total	19·22	19·56	20·30	20·60
Est. value per weaner	21·05	20·92	21·24	20·93
Margin per weaner	1·83	1·36	0·94	0·33
Margin per sow in herd	37·70	26·79	17·11	5·64
Return on capital (excl. interest)	7·9%	5·6%	3·8%	1·3%

Source: Ridgeon, R. F. (1981), Cambridge University, Department of Land Economy.

However, care must be exercised in interpreting the results of such surveys. It is likely that many of the units in the samples practising earlier weaning are more modern units which have incorporated up-to-date knowledge and technology to achieve a more efficient design. It is likely that some, at least, of these units weaning early are operated by more go-ahead and efficient personnel at both stockperson and management level. Thus, it is likely that a large component of the differences in both physical and financial results shown up between different ages at weaning are merely reflecting these differences in application of modern technology and in management ability and stockmanship. It can be argued that if this improved technology and management ability were applied to later-weaning herds, these would be shown up in a much better financial light.

It is important that those pig-keepers practising four to seven week weaning keep this in mind. They may be giving serious thought to the possibility of earlier weaning but it is possible, and even likely, that they would benefit more by improving the efficiency of their present system nearer the 100 per cent mark before embarking on earlier weaning. Those on four to seven week weaning who are already operating near 100 per cent efficiency, and whose buildings are sound, are likely to find it more cost effective to carry out slight modifications to their existing accommodation and system to allow them to wean one to two weeks earlier, rather than, say, reduce from six- to three-week weaning, as this would almost certainly necessitate very substantial capital investment in order to cater adequately for a three-week-old weaned pig.

A crucial consideration in deciding the most desirable age at weaning in any situation is the availability and cost of feed ingredients suitable for inclusion in diets for early-weaned pigs. This applies in particular to dried skim milk. In some countries, dried skim milk is in very short supply or may be available only at excessively high prices. In such situations, age at weaning must be delayed until the maturing digestive system of the pig can digest adequately the type of feed ingredients available for weaning diets.

CONCLUSIONS

On most units practising four to seven week weaning, there is usually adequate scope to increase output from the existing system through management improvements and husbandry modifications before a change to much earlier weaning should be contemplated. For those who are already very efficient on four to seven week weaning, only slight modifications may be required to allow pigs to be weaned one to two weeks earlier. Where buildings are inefficient and dilapidated or where a completely new unit is being planned, and dried milk products are in plentiful supply and available at a reasonable cost, then a strong case can be made from the points of view of sow output and return on capital to base such a unit on earlier weaning. While, at present, there are points in favour of planning a new unit on the basis of either two, three, four or five week weaning, in the UK, three to four week weaning, on balance, is to be preferred.

Chapter 14

MY UNIT—ESTABLISHMENT, OPERATION AND RESULTS

Alastair MacLean

MY AIM in this chapter is to outline briefly what motivated me into pig production, to comment on my basic philosophy regarding the application of sound basic husbandry and well-proven scientific principles in practice, to describe the establishment and operation of my pig unit and to give some indication of the results achieved to date.

After many years specialising in pig husbandry with the National Advisory Service of the Ministry of Agriculture, I was disappointed to note the very small degree of improvement in the output and efficiency of our commercial British pig herd. This situation arose even although there was available a mass of useful information which, if applied, gave promise of resulting in improvements in output and efficiency in our pig herds. There was also available a large number of competent advisers from various organisations, whose prime function was to convey this information to the practical situation.

Was the reason for the failure the result of the 'Boffins' giving impractical guidance? Was it the inability or unwillingness of our pig farmers to take up and develop new ideas and systems, or was the traditional, recurring lack of confidence in the pig industry a deterrent to mass reorganisation and rethinking of our pig enterprises and our traditional approach to pig production?

The reasons were not clear to me, and I began to question whether, in fact, we may have been pitching our national pig output targets at far too high a level.

DECISION TO TEST PRINCIPLES IN PRACTICE

It was in 1972 that I made the decision to attempt to put into practice the advisory principles which I had very forcibly put forward on many occasions to practical pig-keepers and stockmen. I returned to my native Scotland and purchased 'The Meadows' in Aberdeenshire.

THE MEADOWS 200-SOW UNIT

The Meadows was a small arable farm situated, by my choice, well away from existing intensive pig developments, but sufficiently near to marketing outlets. Pigs had not been kept on the farm previously. The amount and type of land was chosen to allow me to deposit the total effluent output from the pig unit, so as in no way to interfere with a normal cropping rotation.

Financial implications were obviously very important to me, and, after repeated cash-flow considerations, my decision was to develop a 200-sow unit producing finished pigs for the bacon market. I was never a believer in the half-way house, and any long-term projections which I had ever carried out left me to conclude that if you commence the job, every possible effort should be made to finish it.

STAFFING

When my unit was producing to its capacity it was operated by two men, with a watching brief and occasional help from myself. I was soon to learn that although initial finance imposed a restriction on size, the two-man 200-sow unit was the wrong size from the point of view of operation, especially with regard to labour deployment. While two men were quite capable of running the unit, there were many factors which interfered with the smooth running of the operation.

The major factor of influence arose in the event of one man being off work through illness, on holiday, or on resignation. The total labour force was then reduced by fifty per cent. It follows, therefore, in my opinion, that the optimum size of herd should be 250 sows (or multiples thereof) run by three full-time men. It must, of course, be understood that the latter suggestion would very much depend upon the layout of the unit, its mode of operation and degree of automation.

When I was involved in providing advice to pig-keepers I always made the point when designing a new pig unit, never to inhibit the possibility of any future extension of the business. In other words, design should be such that expansion should be possible without upsetting the pig flow, and labour utilisation; e.g. all the dry sows should be in one area, etc. The application of this advice into practice in my own situation was very valuable and allowed me to expand and rectify the labour difficulty detailed in the previous paragraph.

ESTABLISHING THE UNIT AND THE SYSTEM

In the initial planning of my unit, I had to take into account the following important implications:

- Finance available.
- Choice of buildings and design.
- Age at weaning.
- Disease control.
- Housing, management and subsequent rearing and finishing.

Finance and size of unit

The finance availability, as mentioned previously, gave rise to the size of unit which was initially built. It is important, at this stage, to get your priorities right, because for example, if I had followed tradition I could well have accommodated only half the number of pigs for the same investment. My decision, therefore, was to intensify as much as possible, making maximum use of floor area. This led me to erect as few farrowing pens as possible because of their expense, which in turn dictated an early-weaning system. I also decided upon floor feeding in the rearing and fattening houses in order to capitalise on high stocking density.

Choice of buildings and design

For reasons of speed, economics and efficiency, a balance was chosen between the erection of package-deal houses and farm-built structures. It is easy to accept that when one knows that the farrowing area is the most expensive to construct on a per sow basis, then the earlier one can wean, the greater will be the reduction in overall building costs. This decision to early wean must, of course, be balanced with the type of structure to be erected for the post-weaning period.

Age at weaning

My decision centred on four-week weaning because I felt that, economically and practically, this was the lowest age to which one can satisfactorily go, for the following reason. All producers know that peaks in numbers farrowing can occur from time to time, and it may be necessary to wean earlier than usual on these occasions. My thinking therefore was that if at any time I had to reduce my weaning age from four weeks to three weeks, then this was likely to impose no great problem. If, however, it had been my policy to wean at three weeks, I would have been unable to wean earlier, thus leaving me no leeway for irregularities of production.

Once one has decided the age at weaning, it then becomes a simple matter to calculate pen size requirement, the design of the unit and the number of the respective buildings required.

Disease control

There would appear to be a number of advantages in deciding upon stock of high health status. The advantages can be found in superior growth rate, lower mortality and morbidity. Disadvantages may lie in the risk of breaking down, and in the smaller number of sources for replacement stock. I decided to opt for breeding stock with a high health status (MD) and after ten years we are happy to remain in this position. It is important when making this decision that absolute security is provided for the pigs. The unit should preferably be surrounded by a wall, and contact with outside risks such as vehicles and people must be avoided.

Housing and management

In looking at national statistics, it struck me that the main area in which breeders were falling down was in the output of litters per sow per annum. It was my plan therefore to create a separate weaning/service building. This was to be the most important part of my unit. It was to be the only area where almost complete free choice of environment would be given to the newly-weaned sow. She would be bedded on straw, whereas at all other stages of production the absence of straw would be total. She would be housed in neighbouring proximity to the boars, and grouped with her batch-weaned mates. I find it advisable to confine the newly-weaned sows with a boar for the first twenty-four hours after weaning. This appears to settle them down more quickly, and less aggression occurs.

The newly weaned sows have access to outside boar/sow 'chatting' areas, separated by tubular steel gates. The inside pens have solid dividing walls to a height of 1·1 metres with open air space above, to allow for free movement of the smell and sound of the boar. The weaned sows have therefore the choice of going outside to 'chat' the flanking boars, or, if they so desire, they can lie peacefully in their straw pen savouring only the stimulating boar smell and noise without face-to-face contact.

POST-WEANING AND SERVICE MANAGEMENT

Serving commences four days after weaning, each sow being brought to her mate, and double-served with a different boar at least twelve hours after her first service, but no longer than eighteen hours later. If these intervals are not achieved, a third service is given. A service is not relied upon as being 100 per cent satisfactory unless complete co-operation between sow and boar is observed, and without protest from the sow. The head stockman assists with all services, each service being visually observed for its duration.

Plate No. **44**

Newly-weaned sows are group housed and deeply bedded on straw. This holiday from concrete floors and individual housing lasts until just over three weeks after service. A boar is put in with the sow for just twenty-four hours after weaning to help minimise fighting among sows.

It is interesting to note at this stage that it is not infrequent for a sow to refuse one boar and readily accept another. It is therefore my view that if a female appears to be on heat, she should have the opportunity to be tried with different boars, preferably of varying size.

The served sows are kept in the service area for a further three weeks after serving, being moved up one pen each week until they are in the proximity of fresh boars at the time of the three-week check. Care must be taken, however, to check served sows daily while in the service house. On a number of occasions I have found sows in heat only one or two weeks after their first service.

It is my belief that in the design of this house we have managed to create conditions which are likely to stimulate the breeding sow.

It provides environmental choice, coupled with group housing, and last but not least, individual attention. The success of the system is confirmed by the fact that, except on the rarest of occasions, all sows are double-served by the morning of the sixth day after weaning, and returns to first service are only 1 per cent, i.e. 99 per cent of sows have conceived to first service.

Feeding in the service area is kept at a low level, with the exception of a 'flushing' boost from one day after weaning until the sow has been served. Liberal feeding during this period is now achieved by allowing sows to feed to appetite from a self-feeding hopper until most in the group have been served. The provision of feed to appetite in this way for a short period after weaning has been found to result in a marked reduction in aggressiveness between sows. The principles underlying the practice of liberal feeding from weaning to mating followed by a reversion to normal feed levels immediately after service are fully explained in Chapters 10 and 11. For any sow which is still on the thin side by service, we bring any required condition back on to such sows after they have returned to their tether stalls some four weeks after service, when we have been known to feed up to 4·5 kg per day for a short period to individual sows.

I have chosen tether stalls for two main reasons:

- They are less costly than totally enclosed stalls and can be accommodated in a much narrower building.
- They allow much better access to the rear of the sow, which allows easier examination for returns to heat or other possible health irregularities.

After ten years of use, I am entirely satisfied with sow tether stalls. The health and well-being of my sows have been excellent. This is reflected in the fact that the stage at which sows are culled from my herd is after 10·6 litters, on average. Many sows in my herd, in fact, are still performing very well up to their seventeenth litter. It is always difficult to make a decision about the culling of an old sow that is still fit and healthy and performing well. In order to reduce the risk of keeping old sows for one litter too many, I am thinking of adopting the practice of culling routinely after a maximum twelve litters regardless of how well sows are performing at that stage. Such a policy would be modified, of course, if the value of the cull sow was reduced relative to the cost of an incoming gilt.

I do not favour bringing boars into the sow-stall house because I believe much more peace is maintained in their absence. Checks are made for returns to service on individual sows at three, six and nine weeks after mating, and all sows are checked each day for returns which may occur out of sequence. In addition, in the last five years, use has been made of the Doppler system of pregnancy diagnosis. This is a useful tool which can confirm pregnancy or barrenness with reasonable accuracy some twenty-five days after service. Doubtful cases must be checked with the machine at a later date. In addition, however, we regularly check the visual appearance of our sows so as to detect any barren animals at as early a stage as possible. Pregnant sows are transferred to the farrowing house at least five days before the recorded 'due-to-farrow' date.

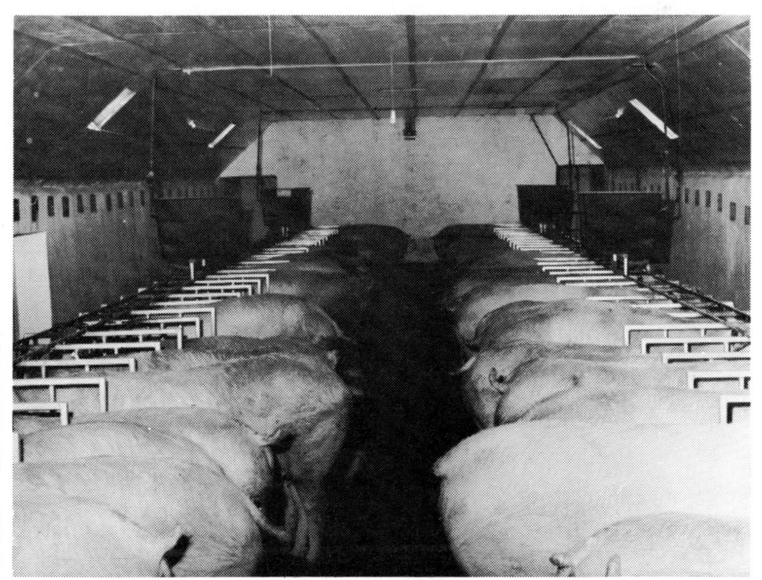

Plate No. **45**

Sows are tethered in stalls from three to four weeks after service.

FARROWING AND REARING

The farrowing houses operate on the room principle on an 'all in all out' basis, and batch weaning takes place. Welded metal flooring for 70 per cent of the pen has given good results, whereas when expanded metal floors were used feet and teat damage resulted. It is interesting to record that, after ten years, I have not had to renew one sheet of welded mesh in the farrowing house although an odd sheet has had to be replaced only recently in the

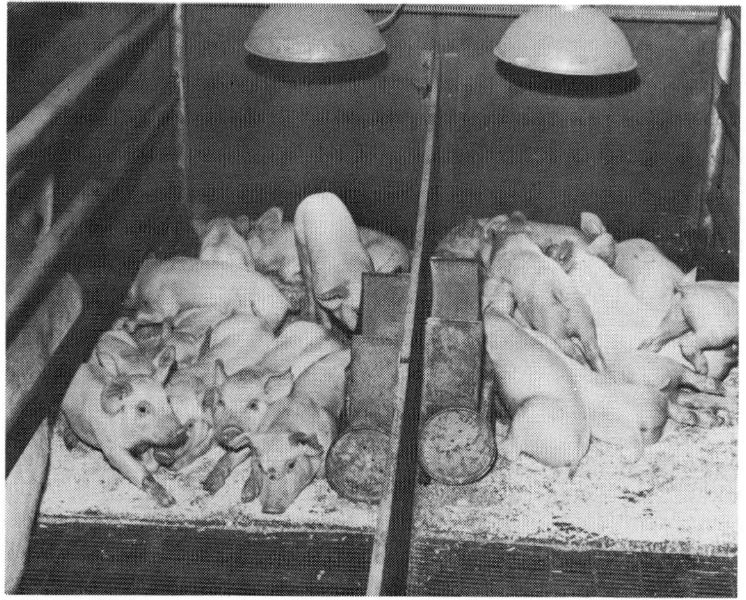

Plate No. **46**

The wide creeps in neighbouring pens are adjacent so as to form a concentrated source of heat in the piglet lying area.

weaner house. The farrowing houses are thoroughly pressure-washed and disinfected between batches.

Almost all farrowings are supervised and cross-fostering is a prime task in order to even up numbers and birthweights, and also to give a further chance to slightly older piglets which may have fixed on to poor milk supply teats. Weakly born pigs are fed by means of a 10 cc syringe, firstly with colostrum removed from newly-farrowed sows, and, when this is not available, we use cow colostrum which has been frozen in blocks until the need to use it arises. It is estimated that through using this technique we save over 50 per cent of pigs which would otherwise be lost. It is just a question of leaving such weakly pigs on the sow and feeding them a few times in the first day or two in this way until they have the strength and confidence to fend for themselves.

While we do not induce farrowings on a routine basis using analogues of Prostaglandin F2α, we do use this approach when the occasion arises. This is most often when we have a comparative shortage of pigs in newly-farrowed sows. In these circumstances, we will induce farrowing one to two days early in sows which appear to be carrying big litters so that surplus pigs can be fostered on to the previous farrowings to make their numbers up to a reasonable level. We are pleased with the results obtained from this careful use of induced farrowing.

On the day of weaning any pigs which do not achieve 5·5 kg live-weight are fostering back to a sow due to be weaned the following week, and a pig in excess of 5·5 kg removed from her litter to leave a suckling space for the smaller pig. This practice has met with complete success.

WEANING

On weaning day, sows are transferred to the service area, and all weaners are removed to a verandah-type weaner house. Ventilation in the verandah house has been modified by me, and differs from the traditional verandah house. In essence it is a higher house, with interior hopper windows above the inside kennels, and it has a closed ridge. This gives me greater control over ventilation compared to a permanently open ridge system.

The verandah house is an extremely tricky one to manage. The correct stocking density of each pen is critical to its success. A good rule of thumb is 9 to 12 kg of pig per 0·1 square metre of lying area plus dunging area. Understocking is more likely to give rise to bad performance than is overcrowding, especially in the first two weeks. I find no problem in adjusting numbers of pigs from pen to pen daily, or as necessary. Fighting never seems to take place, and overall performance figures are good.

In the selection of any type of rearing house, it is important to link up the number of sows which you propose to wean each week to the size and dimensions of the weaner pen. There would be little point providing weaner pens capable of accommodating seventy-plus pigs if the system was based on four sows weaned each week. The outcome would be disaster quite apart from the poor utilisation of space which would very readily be apparent. Understocking generally results in cold pens and fouling of the sleeping area. This combination of adverse factors leads to unthriftiness in the newly-weaned pigs, giving rise to higher mortality figures.

Feeding in the post-weaning stage must be carefully monitored. It is sufficient to feed only small quantities per day during the first four or five days, gradually increasing this to appetite. I have found that the danger period for digestive upsets occurs from eight to twelve days after weaning, and it may be wise to consider feed restriction during this period.

NEED TO REVISE PERFORMANCE TARGETS

My experiences over the past ten years have convinced me that targets higher than those set by the many research, public and private advisory bodies of the UK can be exceeded by far. This will be achieved by paying attention to the detail which I have outlined, *viz* good service procedure, systematic checking, good farrowing house management and cross-fostering with building layout and design tailor-made to the objectives which you have set for the present, not forgetting future expansion possibilities.

The results which I have achieved since the commencement of the enterprise have always satisfied me, in that they have been above the targets which I set myself. Currently, (May 1982) and for the past twelve monthly period, I have been producing 2·43 litters per sow per annum, and we have weaned 10·4 pigs per litter. Monthly pre-weaning mortality (of livebirths) has varied between 1 per cent and 8 per cent, averaging 6·5 per cent. These figures have given me an average output of 25·3 pigs per sow each year.

Table 14·1. Details of diets used—Meadows Pig Unit

Diet	*Energy (Megajoules digestible energy per kg)*	*Crude Protein (%)*	*Lysine (%)*
Sow nuts	12·5	14	0·7
Creep pellets	17·0	18·5	1·0
Grower pellets	13·7	17·5	0·85

Figure 14·1 is a reproduction of one of our standard record cards for individual sows and contains litter details of a sow whose performance was just about average for our herd (25·5 weaners per year). The performance of other sows in the herd has ranged from 19·5 up to almost 30 weaners per year.

Our pigs are moved from the weaner house to the bacon house at 30–32 kg liveweight. This is effected when the pigs are averaging eighty-one days of age, and food consumption has been 78 kg per pig, this figure including sow and boar food, creep feed and grower pellets.

We would like to think that we could improve on these figures, but we have not so far been able to do so. Every man is trained to carry out his own specialised job to the best of his ability, and to understand why he is doing it. It is becoming increasingly difficult to employ dedicated staff, and the success of your pig unit depends very much on the willingness of staff to understand and put into practice the useful research work which has been carried out in this country and abroad.

RECORDING

The recording of one's efforts is a most useful and necessary tool. However, it must not be overdone. Nothing sickens a pigman more than laborious recording to no purpose. It is my opinion that the simplest form of records are all that is required. These should include a sow service book, in which is recorded the date and time of each service, boar used, the three-, six- and nine-week check dates, and the expected date of farrowing. We also record in this book opposite the particular sow, any irregularities at the time of service, e.g. blood from vulva, or protest serve or poor first service, etc. Each sow should have a record card, which stays with her during her lifetime (see Figure 14·1), details being extracted from this in the farm office at each weaning.

The monthly food usage should be recorded accurately. This not only supplies the necessary record of amount of food used per month per pig sold, but also serves as a management tool to arrest any over-or under-feeding which may be taking place on any one section of the unit.

SOW BREEDING RECORD

Sow No	Breed	Date of Birth	Index Score	Number Teats	Sire	Dam
65	COMMERCIAL HYBRID					

Litter No	First Service Date	Boar	Date Farrowed	No Born Alive	No Born Dead	No Weaned	Age Weaned Days	Weight Weaned	Iron Given	Remarks
1	5 9 79		1 1 80	10		10			6 1 80	
2	2 2 80		29 5 80	9		10			5 6 80	2 FOSTERED FROM 62 1 DIED (WEAK AT BIRTH)
3	6 7 80		31 10 80	13	1	11			6 11 80	2 FOSTERED FROM 128
4	2 12 80		26 3 81	12		11			1 4 81	2 FOSTERED FROM 145 2 FOSTERED TO 64 1 CRUSHED
5	4 5 81		26 8 81	11	2	11			2 9 81	1 DEFORMED - DIED 1 FOSTERED FROM 61
6	3 10 81		26 1 82	12	1	10			1 2 82	1 DIED (WEAK AT BIRTH) 1 FOSTERED TO 145
7										
8										
9										
			◄ Totals	67	4	63				
			◄ Mean			10·5				

Date Culled ► [illegible] 2 82

Weaned per Annum ► 25 5

Mortality of Livebirths ► 6%

Reason for Culling ► Low piglet birthweight

Figure 14·1 Copy of the record card of a sow whose productivity was about average for the Meadows herd.

TABLE 14·2. Meadows Herd Statistics 1982

Average number of sows in herd	197	
Age at first service (days)	195	(approx)
Number per litter:		
Stillborn	0·8	
Liveborn	11·2	
Weaned	10·4	
Preweaning losses of livebirths (per cent)	7·1	
Post weaning losses (per cent)	0·5	
Age at weaning (days)	28	(Range 21 to 32 days)
Weaning weight (kg)	7 to 8	
Days to bacon	167	
Weaning to service interval (days)	5	(Range 4 to 6 days)
Returns to first service (per cent)	1	
Litters per sow per year	2·44	
Weaners per sow per year	25·4	
Food intake per 30 kg weaner (kg)	78	
Includes share of sow and boar food, creep and grower pellets)		
Average number of services per boar per week	3	
Average number of litters per sow before disposal	10·6	

FUTURE DEVELOPMENTS

I am sure that our future must lie in further expansion, if only to increase the opportunity for specialisation of labour and for motivation of staff through a healthy competitive element between colleagues.

On the other hand, the ups and downs of the pig industry over the past decades do not stimulate the confidence necessary to increase the size of one's business and there is no doubt in my mind that pig production in the UK cannot survive if we are to operate at present low average levels of output.

Low output means low efficiency, which at the end of the day results in an overpriced product which may be unsaleable.

Chapter 15

IMPROVING EFFICIENCY—MANAGEMENT AND LABOUR CONSIDERATIONS

It is good business to get more out of an existing resource while avoiding an appreciable increase in costs. As indicated in the first chapter, most of the costs in weaner production are incurred whether the sow rears ten or twenty-six weaners per sow per year, and, this being so, it is only common sense that every effort should be made to maximise the output of weaners from each sow in a year.

PERFORMANCE IN RELATION TO THEORETICAL TARGETS

The Meat and Livestock Commission (MLC) in the UK publish a summary of results from their Pig Plan Recording and Costing Scheme in the form of averages of all herds and the average of the top 10 per cent of herds. Selection of the top 10 per cent of herds is based on weaners produced per sow per year. The most recent results from this scheme which, if anything, may tend to involve the better-managed herds, are presented in Table 15·1 and the theoretically possible performance is inserted for comparison. Although the data is analysed separately for herds weaning at 10–18, 19–25, 26–32, 33–39 and 40–63 days of age, Table 15·1 contains only the data for 19–25 day weaning since this comprises the biggest sample of herds (142 herds in total).

TABLE 15·1. Performance of pig herds weaning at 19–25 (mean 22) days participating in the MLC Pig Plan Recording and Costing Scheme (1980–81)

	Average	*Average of top 10 per cent*	*Theoretically possible target*
Litters per year	2·31	2·39	2·55
Born alive per litter	10·2	11·0	11·5
Losses of livebirths (%)	10·6	10·1	5·0
Weaned per litter	9·12	9·89	10.9
Reared per sow per year	21·1	23·6	27·8

Thus, while the top 10 per cent of herds perform much better than the average in most aspects of performance, even these top herds fall appreciably short of the target which is theoretically possible and it appears that only a handful of herds succeed in achieving this target. Of course, the knowledge that herds differ in output and efficiency is one thing, but of much greater importance in relation to determining how best to increase efficiency is the reason for these differences between herds, and often the exact reasons are far from obvious.

Within even the average or below-average herds there are likely to be some individual sows which attain the target performance level of almost twenty-eight weaners per sow per year on three- to four-week weaning.

Thus, the theoretical target is achieved by only a handful of herds although a few individual sows in many other herds will be achieving this target. At the other extreme, of course, many herds will be performing well below the average, while many sows in individual herds are likely to be achieving extremely low levels of output.

DETECTING INDIVIDUAL SOW AND HERD DEFICIENCIES

So in setting out to improve the efficiency of sow output, this must be tackled not only by examining both the herd in general and its management, but also by scrutinising carefully the output of all individuals within the herd.

Regarding the individuals within the herd, the performance of individual sows must be monitored carefully and frequently so that problems can be detected promptly leading to appropriate remedial treatment or timely culling. It must be remembered, of course, that while each individual sow has an important influence on herd productivity and efficiency, the individual boar has even a greater influence and hence his performance should be monitored even more carefully and frequently. An efficient recording and monitoring system for each sow and boar in the herd is an essential basis for detecting and dealing promptly with 'passengers' or the inefficient individuals within the herd.

The main 'herd' factors to be examined critically in setting out to improve the efficiency of sow output include:

- The soundness and efficiency of the stock from the genetic standpoint.
- The health of the stock.

- The facilities, i.e. buildings, environment, feeding arrangements etc.
- The system of operation.
- The management ability.
- The ability of the stockmen or women.
- The organisation and motivation of labour.
- The time available to management and staff to exercise their skills.

So far this book has dealt with methods for improving the genetic merit and efficiency of breeding stock, of improving or maintaining sound health and of providing both adult and young stock with housing, climatic conditions and feeding appropriate to their needs.

The one major component for achieving a high level of efficiency which has so far not been discussed at any length is management and stockmanship.

MANAGEMENT

It is the role of management to make available to stockmen genetically efficient and healthy stock, good facilities, a sound system and to provide the stockmen with the necessary motivation and time so that they are given every incentive and opportunity to exercise their skills.

LABOUR MOTIVATION AND DEPLOYMENT

It is much easier to motivate and deploy labour effectively if the objectives of their job are well defined. In setting out to achieve a high level of output and efficiency in weaner production the objectives are very clear cut.

What is basically required are high numbers of viable piglets born per litter, low pre-weaning mortality, minimum weaning-to-conception interval and minimum 'empty' or unproductive days for the herd in general. However, in setting out to achieve these straightforward objectives there is no single magical formula for success. Success in achieving each of these objectives depends on applying a multiplicity of small factors.

Thus, in order to achieve a high level of performance, there must be full appreciation by the stockmen of the factors involved, a system incorporating these factors must be developed carefully, preferably through the joint efforts of management and staff, there must be implicit belief in this system on the part of the stockman who must then apply it rigorously in practice.

In order to apply the system properly, the stockman should be allocated sufficient time to pay the necessary attention to detail. The philosophy of getting one person to look after more and more sows has proved to be false economy on many units because, in this situation, there is increasingly less time to pay the necessary attention to detail. Performance per sow consequently suffers.

Personnel with traditional, preconceived ideas which may be less than half-truths are often difficult to motivate in an alternative but well-proven direction. Consequently, intelligent and energetic personnel entering the modern pig industry without any preconceived ideas often perform excellently, given the proper guidance and training.

Careful recording and monitoring of the pig herd in both physical and financial terms and the involvement of all pig unit personnel in regular discussion on the trends in performance and on possible reasons for failure to achieve target performance levels can be an extremely useful exercise in terms of providing adequate motivation.

STOCKMANSHIP

Of all the vital components for achieving a high level of efficiency in weaner production, stockmanship is undoubtedly the most important. The leading scientists who have grappled with the requirements of the pig in terms of its nutrition, housing and maintenance of its health are the servants; the good stockman is the master. The buck has to stop somewhere and it is the good stockman who, at the point of practical application, has the responsibility for moulding existing knowledge on disease prevention, feeding, housing and provision of an adequate environment into a system of operation in the actual commercial production of weaners. The behaviour and performance of the stock act jointly as his main barometer of well-being and he adapts and amends the basic system according to the needs of individual animals within the herd.

Stockmanship is one of those terms which is extremely difficult to define. It is a well-moulded combination of a sound basic knowledge of the subject, a basic attachment for and patience with the stock, an ability to recognise all individual animals and to remember their particular eccentricities, a keen sensitivity for recognising the slightest departure from normal behaviour of individual animals, an ability to organise his working time well, a keen appreciation of priorities combined with an almost constant willingness to be side-tracked from routine duties as pressing needs arise to attend to individual animals in most need of attention. The basic desire is to make each animal as comfortable and contented as possible and to strive constantly for higher levels of performance and efficiency.

Stockmanship, being such a vital component in achieving high efficiency in weaner production, creates a pressing need for ensuring a good flow of suitable personnel into the industry and the appropriate training of such is brought into focus. Personnel cannot be trained in a College classroom, and training is best obtained through an apprenticeship working with stock under the guidance of a patient, knowledgeable and experienced stockman with the ability to train others in his skills and to impart to them the secrets for his success.

The value of skilled stockmanship has always been recognised by those who knew how to get the best out of their stock. In this age of automation, stockmanship has not become less important, but even more so, if we are to get the most out of expensive stock, buildings and feed. Good stock and a sound basic system can very readily end up in disaster in the hands of inadequate management and labour. Good stockmanship, on the other hand, can be looked upon as a vital cog in a wheel which gets the very best out of mediocre stock and a mediocre system and makes sure that a potentially good system fulfils its full promise in terms of efficient weaner production.

CONCLUSION

The pig industry is generally inefficient. It is probably no more so than the sheep, beef or dairy industry in this respect but this is no consolation. With average margins likely to decline in the future because of recurring crisis associated with the pig cycle and uncertainty over the price of pigmeat compared to the cost of feedingstuffs, all pig producers who earnestly wish their pig business to be profitable will be forced to revolutionise their attitude, their system and their methods of operation and they will have to set their sights higher so as to achieve considerably increased output and efficiency.

Fortunately, we already have a handful of very efficient producers. They have shown what can be done through setting high objectives, wise planning, applying sound principles and efficient operation. They are well prepared to meet the challenge of the future. The challenge to all other pig producers is to follow the leaders.

Chapter 16

WELFARE OF THE SOW AND HER PIGLETS

A WIDE VARIETY of pig systems exist in the UK to fit in with the farming pattern and climate. The great majority of these systems exist for sound reasons and therefore sound justification must be found for imposing restrictions on the range of systems available.

PIG-KEEPING SYSTEMS

Some of the more recently developed systems are being criticised on the grounds of welfare. These include, in particular, the use of individual stalls and tether systems for dry sows and cage-rearing systems for weaned pigs.

Criticisms of animal welfare come from two main sources, as cited by Lindgren in 1976:

1. those individuals with a very genuine interest in animal welfare; and
2. individuals or organisations who use the animal welfare issue for politcal, media or radical objectives. Lindgren claimed that this latter group had little regard for the truth and were not swayed by it.

The systems of animal production which are currently being criticised were developed with the inseparable objectives of improving aspects of welfare, animal productivity and business efficiency when compared with previously existing practices. The animal production industry has an outstanding record of improving efficiency, and so keeping down the cost of animal products to the housewife. Reverting to alternative systems might involve higher production costs and the limited evidence available suggests that only a small proportion of the consuming public would be prepared to pay the higher prices for their food required to cover the higher production costs involved.

Balance of advantages and disadvantages of each system

All concerned with animal production systems would readily admit that no system is perfect. All systems, whether intensive, semi-intensive or extensive have their respective advantages and disadvantages. The relative deficiencies and advantages of a particular system cannot be gauged by the superficial examination of that system at a favourable time of the day or of the year; nor can the system as a whole be criticised or praised on the assessment of only a few examples of that system. There is a wide disparity between results from different examples of the same system on different farms when measurements are made of pig behaviour, apparent comfort, contentment and productivity. Increased knowledge of the reasons for such variation would be of great help in improving all systems of animal production.

Variation between examples of the same system

The considerable variation noticeable between different examples of the same system on different farms could be due to the different stages in the development of the system on these farms. The more recent systems of pig production, for example, are still at a comparatively early stage of development and an optimal model of each system may not yet have evolved or may have been developed on only a very small proportion of farms. Thus, it is likely that many of the newer pig production systems have been prematurely, and therefore unfairly, judged. In other industries, the final judgment on prototypes is not carried out on the early models but at the final stage of development, after problems isolated in the earlier prototypes have been solved and the systems continuously improved and modified until the final version is produced.

Accordingly, for some of the pig production systems which have been subject to criticism, suggestions are made below on how these systems can be improved to accommodate the welfare dimension and improve the efficiency of the system at the same time. In addition, suggestions are made for the improvement of some 'traditional' systems.

Critical evaluation and improvement of existing systems

Systems for sows

The main alternative systems for housing newly-weaned and pregnant sows are as follows:

(a) Groups of sows in strawyards.
(b) Small groups of sows in pens or kennels with adjacent dunging area and with or without individual feeding stalls.
(c) Small groups of sows in the cubicle system.

(d) Individual stall and tether systems.
(e) Small groups of sows with free access stalls.
(f) Individual pens.

In this chapter, the main discussion will focus on the extremes of these systems, that is, strawyards and individual stall and tether systems. Very little comparative data on aspects relating to welfare are available on these alternative systems. Carter and English (Unpublished) have recently carried out some relevant pilot studies, the results of which are interesting. A total of nine commercial sow housing systems were examined on different farms in Aberdeenshire between August and October 1982. These comprised three strawyards, two concrete-floored kennels with outside yards, two stall systems and two tether systems. Activity and behavioural aspects of sows were recorded in each system on two consecutive days during the hours 06.00–11.00, 13.00–18.00 and 20.00–21.00. All sow groups examined were well established. The incidence of aggressive encounters observed is detailed in Table 16·1.

As might be expected from the nature of the housing system, very few aggressive encounters were noted in the stall and tether systems while such encounters were numerous in all group systems. Aggressive activity was associated with competition for food, water and apparently for lying areas. Only one watering point was available to each group in the group housing systems and this appeared to contribute to much of the aggression observed on at least one farm.

While aggressive encounters were almost negligible in stall and tether systems, bar and chain biting was evident in these systems, but the incidence varied from being considerable on one farm to being almost negligible on another. The time the sows spent lying was similar in strawyards and tether systems. Sows spent most time lying when in stalls. The number of postural changes was similar for sows on strawyards and stall systems, but was less in tethers.

One can draw few conclusions about the relative problems of different systems from this small preliminary study other than to lend emphasis to what is already generally appreciated:

1. There is considerable aggression between sows even in established groups, especially when feeding, drinking or in choosing a resting site.
2. The incidence of stereotypic behaviour such as bar and chain 'biting' or 'chewing' in stall and tether systems varies widely between farms.

TABLE 16·1. Relative incidence of aggressive encounters on alternative sow housing systems

System	*Total floor space per sow (m^2)*	*Number of sows in house/pen*	*Number of sows observed*	*Number of aggressive encounters per sow per 11 hour period*	
Strawyard:					
1	2·43	37	37	11·1	
2	2·5	41	41	10·1	mean = 13·1
3	2·4	30 + 1 boar	30 + 1 boar	18·1	
Concrete floored pen and yard:					
1	2·22	8 + 1 boar	8 + 1 boar	6·1	mean = 8·0
2	2·22	8 + 1 boar	8 + 1 boar	10·0	
Stall system:					
1	2·63	208	22	0·6	mean = 0·6
2	2·97	212	20	0·6	
Tethered system:					
1	1.78	542	20	0·6	mean = 1·0
2	1·83	72	20	1·3	

From Carter and English (Unpublished).

Aggression to a greater or lesser degree is always likely to be a feature of group housing systems for sows. Groups of sows can rarely be permanent since a group is subdivided prior to farrowing and a new group reconstituted following weaning. When sows are grouped in this way they will fight to establish their social status and the larger the group the more fighting there will be. The worst effects of aggressiveness might be reduced if more space, individual feeders and more watering points were provided. These extra provisions will obviously increase costs and will not completely eliminate aggression for this occurs at times other than at feeding or drinking. Often the only solution is to house timid and hyper-aggressive sows in individual pens. It was largely to solve the problem of aggression and injury that individual stall and tether systems developed. These latter systems do place severe restrictions on the animal's opportunity for exercise and may cause frustration but, on the other hand, they do ensure security and adequate feed and water for each individual sow.

Other factors which must be taken into account in assessing the relative merits of alternative housing systems for dry sows include the floor space required per sow, housing costs and the ability of each systems to provide an adequate thermal environment for the sow. These aspects as they relate to alternative systems are summarised in Table 16·2.

The system with the lowest capital cost per sow place of those listed in Table 16·2 is simple strawyards, followed by tether stalls. The most expensive systems are free-access stalls and individual pens. When individual feeders are incorporated into a strawyard system, such a system becomes one of the most expensive ones.

Straw-bedded systems provide the highest degree of thermal comfort in cold weather provided a bed of good dry straw is available. When bedding is allowed to become wet, systems using bedding have little or no thermal advantages over systems not using bedding. The greatest heat loss takes place through the ventilation system which makes it very important to provide minimum ventilation in cold-weather conditions.

As well as considering the relative costs and thermal comfort levels on alternative systems, it is important to minimise food costs. Higher food costs are usually incurred on group housing systems without individual feeders and in systems providing lower thermal comfort levels.

Methods of improving housing systems for sows will now be summarised.

TABLE 16·2. Comparison of different housing systems for dry sows

System	Group size	Floor space per sow (m^2)	Building cost per sow (A)		Thermoneutral zone (C)		Minimum outside temperature to allow house to be maintained at or above LCT without supplementary heating	
			£/m^2	Index (B)	LCT (°C)	UCT (°C)	Ventilation rate 0·395 m^3/S (E) (0°C)	Ventilation rate 1·354 m^3/S (F) (0°C)
1. Sow stalls with concrete slats	1	2·42	260	163	21·0	33·0	3·0	12·0
2. Tether stalls with concrete slats	1	2·20	215	134	21·0	33·0	2·0	12·0
3. Strawyards	10	2·34	160	100	12·0(D)	31·0(D)	−5·0	5·0
Strawyards with individual feeders	10	4·21	325	203	12·0(D)	31·0(D)	−2·0	5·0
4. Sow cubicles with bedded floor	5	3·76	250	156	12·0	30·0	−2·0	5·0
5. Free-access stalls with concrete-slatted floors	5	3·76	320	200	21·0	33·0	3·0	12·0
6. Kennels with solid concrete floors + Outside yards with individual feeders	5	2·10* 2·80†	255	159	18·0	33·0	1·0	9·0
7. Individual pens with mesh floors	1	3·64	360	225	21·0	33·0	4·0	12·0

* = inside, † = outside.
(A) Costs at December 1980. Wight, H. J. and Clark, J. J. (1981). *Farm Buildings Cost Guide 1981*, SFBIU Aberdeen.
(B) Strawyards without individual feeders = 100.
(C) LCT, UCT and temperature conditions calculated using models by Bruce and Clark (1979).
(D) With very wet poor straw conditions LCT = 16°C, UCT = 33°C.
(E) Ventilation rate estimated at minimum = 0·3 per cent CO_2 concentration of ambient air.
(F) Generally recommended level.
LCT = Lower critical temperature. UCT = Upper critical temperature.
Calculations of LCT and UCT based on a daily intake of 2·2 kg of a diet with 12·7 MJ DE/kg.
Figures for free-access stalls based on assumption that sows lie mainly in stalls rather than as a group in the exercise area.

Improvement of housing systems for dry sows

On the basis of the limited amount of information available, it is suggested that the basic provisions which should be made in alternative dry sow housing systems to cater for the adequate well-being of the animal are as follows.

Group housing systems

1. Avoid large groups of sows (preferably not more than ten sows per group evenly matched for age and weight if possible).
2. Instal individual feeders to reduce competition and to facilitate feeding according to body condition.
3. Provide adequate watering facilities to minimise competition (at least two drinking points spaced well apart for every group of sows). Choose drinking devices which reduce spillage and do not block easily with bedding.
4. In poorly-insulated buildings provide kennels.
5. In straw systems, maintain straw bedding in adequate quantities and in dry condition.
6. Provide some individual pens or stalls or tethers to accommodate the timid and/or hyper-aggressive sows and also those sows suffering from malnutrition due to competition.

Individual housing systems (e.g. stalls and tethers)

1. Provide a well-insulated and adequately ventilated building.
2. Keep floors in a good state of repair.
3. Maintain an adequate level of thermal comfort in the building.
4. Slat width should be not less than 100 mm with a maximum 25 mm gap between slats. The best type of slat is one of reinforced concrete with rounded 'pencil' edges.
5. Use a suitable tether which eliminates the risk of abrasions.
6. Minimum width of stall should be 610 mm to ensure that the sow can lie comfortably and experiences no difficulty in either lying down or standing up.
7. Keep sows in good body condition.
8. Arrange watering system so that the risk of the sow's lying area becoming wet is eliminated.

Use of straw in stall and tether systems

The Revised Code of Recommendations for the Welfare of Pigs (1981) produced by the Farm Animal Welfare Council in the UK

(The Draft Welfare Code, 1981) makes a strong recommendation for the use of straw bedding in stall and tether houses to make sows more comfortable and contented. A good bed of dry straw has a thermal equivalent of about 4 to 5°C and does provide a source of recreation for sows.

However, unless properly managed, such bedding could increase drainage problems and, once wet, would reduce comfort levels. This could eventually lead to injuries and painful diseases such as cystitis and mastitis.

Cage-rearing systems for weaned pigs

Both the Draft Pig Welfare Code (1981) and the House of Commons First Report from the Agriculture Committee on Animal Welfare in Poultry, Pig and Veal Calf Production (1981) (House of Commons Committee, 1981) recognised the advantages of accommodating weaned pigs on perforated floors. Minimising the amount of contact between young pigs and their own dung and urine helps to reduce the risk of disease transmission. Both parties, however, considered that part of the floor area should be bedded to improve comfort. There is a need for greater comfort at lower cost in many systems, but the use of bedding is only one alternative option. The use of straw in some systems presents operational difficulties. The provision of a bedded area, for example, could easily contribute to the rapid dissemination of enteritis as well as cause discomfort when wet, the combination of which creates serious welfare problems. The objectives of the

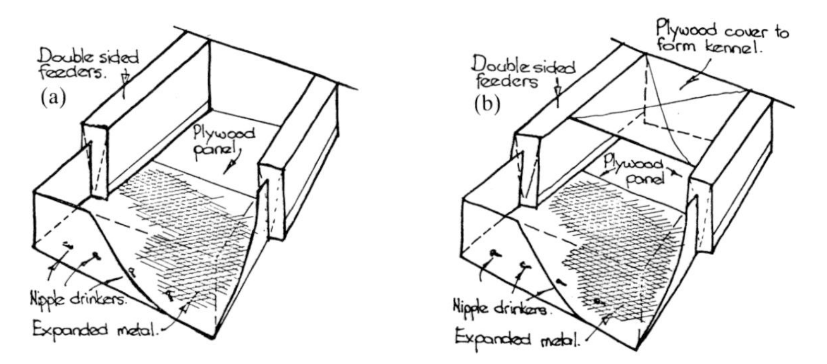

(a) *Flat-deck pen partly covered with removable solid floor above the mesh to provide a lying area.*
(b) *Kennel constructed over removable solid floor.*

Figure 16·1. Modified flat deck pen designed to increase pig comfort and reduce heating costs.

Farm Animal Welfare Council and the House of Commons Committee (1981) towards providing the weaned pig with a higher degree of comfort might be achieved more effectively in practice by providing a limited area of solid floor (e.g. asbestos or sealed wood) above well-designed perforated floors to provide a lying area. This area of solid floor should be removable in the event of an outbreak of scour. This solid area would also serve as the base of a kennel which would give protection against draughts, a high level of comfort and reduce the use of fossil fuel energy to create a suitable thermal environment for early weaned pigs (see Figure 16·1).

Such an arrangement would have the following advantages:

1. Provision of a comfortable lying area for piglets.
2. Reduced need for fossil fuel energy to create a comfortable resting area for early-weaned pigs. (The conservation of the piglets' own metabolic heat production within the kennel will raise temperature by at least 5°C.)
3. It provides pigs with a choice of environment within the pen.

Other features of the arrangements illustrated in Figure 16·1 are:

1. Liberal feeding space and easy access of all pigs to feeders by providing double-sided feeders down each side of the pen. Early-weaned pigs should have a highly digestible diet which can be fed to appetite.
2. Liberal watering facilities (at least one watering point which can be readily operated for every ten pigs with a minimum of two watering points per pen).
3. Where provision of the kennel fails to provide an adequate thermal environment for early-weaned pigs, a heat lamp can be suspended to fit neatly into a circular space in the middle of the lid of the kennel.

The saving in heating costs on the kennel system is partly offset by the fact that inspection of pigs is more difficult than in a conventional flat-deck system. Therefore good discipline is required in kennel systems to ensure regular inspection of pigs. However, the system does provide a choice of environment and more liberal provision of essential requirements which has the effect of keeping newly-weaned pigs more comfortable and contented, a combination of factors which helps to minimise the incidence of abnormal behaviour and vices.

Farrowing pens

Both the House of Commons Committee (1981) and the Farm Animal Welfare Council in the Draft Pig Welfare Code (1981) recognised the advantages of the farrowing pen incorporating a farrowing crate and a heated creep area for the piglets.

However, the trend in farrowing pen design is to provide a higher proportion of slatted area on the floor. A slatted area behind the sow helps to keep the farrowing pen dry and clean but the forward extension in the pen of this slatted area has probably resulted in a reduction in comfort level in many farrowing pens. This may well be a retrograde development from both the comfort and efficiency viewpoints, for a higher proportion of solid floor area, especially at the front of the pen, makes it possible to provide adequate dry bedding, for example wood shavings, particularly during, and for a short period after, farrowing.

Thus a greater area of suitable solid floor area and provision of suitable bedding could usefully be restored to many farrowing pen designs and such provision is likely to improve comfort for newly-born piglets, in particular, and to reduce mortality. Such specialist 'maternity' pens could be used to accommodate the sow and litter until the latter is properly established a few days after farrowing, at which stage they could be transferred to 'farrowing' pens incorporating a greater area of slatted flooring to achieve the twin objectives of a healthy rearing environment and labour economy.

MUTILATIONS PERFORMED ON PIGS

(1) Castration

Castration is the most cruel operation performed on the farm.

Apart from avoiding such cruelty, the advantages of leaving boars entire are as follows:

- Improved growth rate;
- Improved feed conversion efficiency;
- Higher carcase lean content.

The disadvantages of entire boars are:

- Slightly heavier shoulder development;
- Slightly lower killing-out percentage;
- Lower absorption of brine during curing;
- Risk of 'boar taint' in the carcases of boars which were old enough to have reached puberty by slaughter.

When boars are produced for pork and bacon it is essential that they be subjected to a liberal feeding system. This will help to ensure an adequate carcase fat level at slaughter which will help to reduce the deficiencies in boar carcases claimed by meat processors such as excessive leanness and lower uptake of brine during curing and at the same time help to eliminate the risk of boar taint. Boar taint and the welfare aspect apart, it is the case that the balance of considerations are in favour of leaving boars entire.

The incidence of boar taint is negligible in modern pigs slaughtered at young ages at bacon weight (90 kg liveweight) or at lower weights.

Because of the very low incidence of boar taint in pigs slaughtered for bacon or pork at young ages, the House of Commons Committee (1981) were strongly tempted to advocate a complete ban on castration. The only factor which discouraged this recommendation was the fear that it would be difficult to achieve international agreement on such a policy. If castration was banned in the UK and continued in other countries, then the UK might be subject to unfair competition in pig products imported from such countries.

Because of (1) the harsh nature of the operation, (2) the fact that the job is detested by all skilful operators with feeling for the animal, (3) the fact that there are certainly no net economic advantages in continuing this practice and (4) the overwhelming evidence that the incidence of boar taint is negligible in genetically improved pigs grown quickly to slaughter for bacon or pork, it is essential that every effort be made to reach general international agreement as quickly as possible to make the castration of pigs destined for the pork and bacon markets a thing of the past.

(2) Teeth-clipping and tail-docking

The operations of teeth-clipping at birth and tail-docking in the first few days of life must cause temporary discomfort to the piglet. The Draft Pig Welfare Code (1981) suggested that these 'mutilations should be avoided wherever possible'. This recommendation was endorsed by the House of Commons Committee (1981) who further expressed the 'hope that any producer who still docks his piglets' tails as a routine measure will pause and ask himself whether what must be a tiresome and laborious task could not be avoided by modified management'.

The problem associated with the failure to clip piglets' teeth at birth and dock piglets' tails in the first few days of life is that the consequences cannot in any way be predicted. Not all litters will damage each other and the udder of the sow if their needle teeth are not clipped. However, there is no method of determining in advance which litters could be left intact with impunity and which litters must have their needle teeth clipped, if serious scarification of the faces of littermates and of the sow's udder is to be avoided. This being the case, there is absolutely no alternative than for competent trained operators to clip the needle teeth of all piglets at birth as a routine.

The same is the case with tail-docking. In a well-designed cage-rearing unit suitably modified and 'enriched' as already described and operated in an efficient manner, the risk of tail-biting is absolutely minimal. The risk of an outbreak of tail-biting is certainly no higher than it is in a well-designed and well-operated deep-strawed system. An outbreak of tail-biting is often spontaneous and the basic reasons for such an outbreak are extremely difficult, if not impossible, to determine. We have experience of serious outbreaks of tail-biting on apparently well-designed and well-operated deep-strawed systems and we have often found it impossible to isolate the factor or factors responsible.

As it is often impossible to predict spontaneous outbreaks of tail-biting in apparently well-designed and well-operated systems, the only logical approach to preventing the damage and suffering caused during such an outbreak is for a competent, trained operator to dock piglets' tails in the first few days after birth as a routine.

Routine teeth-clipping and tail-docking at or very soon after birth are analogous to vaccination programmes applied to young human infants. Such vaccinations cause temporary discomfort and pain but they serve to prevent much more serious problems in later life which might otherwise affect a very small proportion of individuals.

IMPORTANCE OF GOOD STOCKMANSHIP

The crucial importance of good management and stockmanship for ensuring the adequate welfare of farm animals on any system is universally recognised. The House of Commons Committee (1981) stated that the most important factor affecting the welfare of animals was the standard of knowledgeable, conscientious and sympathetic care bestowed on them. It is generally accepted that good stockmanship has a much greater influence on animal welfare than alternative housing systems. This being so, it is important to ensure adequate training in the science and practice of good stockmanship.

TECHNIQUES FOR ASSESSING WELL-BEING ON IMPROVED SYSTEMS

Many approaches to measuring welfare have been used, the main ones being based on behavioural, physiological and productivity measurements. Physiological and biochemical measurements suffer from the disadvantage that they are not very easy to take, particularly in the field. In addition, many of these measurements entail the slaughter of the animal or involve restraint of the animal which is itself likely to be stressful. Behaviour studies which measure general activity and incidence of abnormal or stereotyped behaviour can provide useful indications of the well-being of the pig on a particular system. However, if such behaviour studies are to be useful they must involve study over representative periods of the day and of the year and must also cover several examples of the same system on different farms. While productivity measurements have been somewhat maligned by some authorities in relation to their usefulness in measuring pig welfare we are not impressed by the reasoning behind their conclusions. It is our view that the level and efficiency of production of the herd, and of the

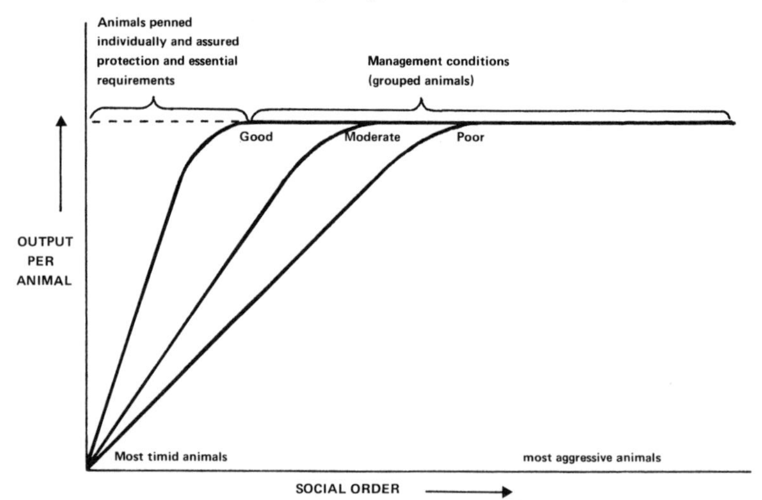

Figure 16·2. Performance of timid animals of low social status kept in groups under good, moderate and poor management conditions and when penned individually. Adapted from McBRIDE (1964)

individual pig in particular, representing as it does the degree of harmony of the animal with its environment, constitutes a reliable measure of welfare in most situations.

This view is fully supported by the work of the behaviourist, Glen McBride in Australia whose poultry model based on actual data is illustrated in modified form in Figure 16·2.

McBride classified laying pens on the basis of their aggressiveness and thereafter monitored their performance within groups subjected to different levels of management (see Figure 16·2). It can be seen that when grouped animals are subjected to 'poor' management, the most aggressive animals do not suffer in terms of performance. However, a high proportion of the less aggressive animals perform poorly, performance declining to very low levels of output in the case of the most timid animals. Progressively better management of grouped animals (improving from 'poor' to 'moderate' to 'good') results in a higher proportion of animals performing to their potential. However, even when grouped animals are subjected to 'good' management, a proportion of individuals perform less well, the most timid animals having very low levels of output. The only way in which McBride could get the most timid animals to express their full potential was to house them in individual pens, thus protecting them from the aggression and competition of their fellows and assuring them of their basic requirements of feed, water and resting space at all times.

The ability to protect the most timid animal in the herd or flock and to ensure the essential requirements of such an animal, is the basic test, in welfare terms, of any animal production system. For, if a given system ensures maximum protection and welfare for the most timid animal, then more aggressive individuals will obviously also be catered for in terms of their basic needs.

Of the dry sow housing systems so far developed, only the well-designed and well-operated individual stall and tether systems ensure protection and adequate basic requirements of food, water and resting area for the most timid sow in the herd at all times.

Grouping of sows certainly has many attractions so far as opportunities for exercise, exploration and social contact are concerned. The challenge in developing the perfect group system is considerable, however, for it must ensure maximum security for the most timid sow in the herd for twenty-four hours per day and for 365 days in the year.

It is likely that the whole debate on the welfare of the sow would be much more rational if the opinions of sows could be determined about the housing/management system that were most acceptable to them when given the choice of alternative well-designed and well-operated systems. However, it is most unlikely that unanimous answers would emerge from such an 'opinion poll'. It is likely that the aggressive sow would express a definite preference for group housing systems, for in such a situation she would stand to obtain more than her fair share of resources at the expense of her more timid pen-mates. On the other hand, the timid sow would almost certainly express preference for a well-designed and well-operated individual stall or tether system because of the security and fair share of resources which such a system ensures.

NEED FOR CONSUMER EDUCATION

Because the great majority of consumers are now urban-based there is a general lack of understanding among them about the challenges farmers have had to face in increasing the efficiency of food production. It is vital that every effort is made to educate the consuming public about modern animal production systems and the reasons for their existence.

CONCLUSIONS

The major systems of pig production in use today have all evolved for very sound reasons. The systems which are currently being criticised were developed with the inseparable objectives of improving aspects of welfare, productivity and business efficiency when compared with previously existing practices. It has to be conceded that more recently developed systems, like their predecessors, possess some disadvantages. However, these deficiencies are recognised by developers and the systems are in a state of dynamic development and further improvement.

'There is no doubt that, at present, improvement of existing systems will yield a greater gain in terms of animal welfare than that which would be obtained by legislating that one system of production is better than another.' (British Society of Animal Production, 1980).

However, pig-keepers have no cause to be complacent. While the majority of pig producers take every care in ensuring that their animals, as far as they can judge, are perfectly warm, comfortable and contented, there are exceptions to this general rule and such people are being very unfair to their animals and at the same time are doing a great disservice to their fellow-producers.

The overriding importance of the efficiency of management and stockmanship imposed on any system of pig production in ensuring adequate standards of pig welfare and productivity cannot be over-emphasised. This being so, it is important to ensure adequate training in the science and practice of good stockmanship.

References

Draft Pig Welfare Code (1981). Draft Revised Code of Recommendations for the Welfare of Pigs. Annex D. MAFF. Tolworth, Surbiton, Surrey KT6 7NF.

House of Commons Committee (1981). House of Commons first Report from the Agriculture Committee. Session 1980–81. *Animal Welfare in Poultry, Pig and Veal Calf Production,* Volume 1, HMSO, London.

Kilgour, R. (1980), *Animal Welfare—The Conflicting Viewpoints,* in *Behaviour in relation to reproduction, management and welfare of farm animals,* Proceedings of a symposium, University of New England, Armidale, NSW, Australia, September 1979. (Ed. M. Wodzicka-Tomaszewska; T. N. Edey, and J. J. Lynch). Reviews in *Rural Science* IV, 175–182.

Lindgren, N. O. (1976), 'The conflict between technical advances and ethics in animal production', *Wld. Poult. Sci. J.* 32, 243–248.

Acknowledgement

Parts of the foregoing chapter are based on the following paper: English, P. R., Baxter, S. H. and Smith, W. J. (1982), 'Accommodating the welfare dimension in future systems', paper presented to the British Society of Animal Production, Harrogate, Yorkshire, March 1982.

ACKNOWLEDGMENTS

ANY 'HUSBANDRY-BASED' book must be founded on knowledge gleaned over many years from research, farm investigations and from the experiences of practical pigkeepers. This book is no exception.

We are indebted to a great many researchers and other professional colleagues for information used in the book. We are particularly indebted to those who agreed to read parts of the manuscript. These include: Mr W. M. Caldwell (Chapter 1); Dr J. T. Done, Dr A. E. Wrathall and Mr L. G. Donald (Chapters 3, 10); Professor J. B. Owen and Dr Maurice Bichard (Chapter 4); Dr P. H. Brooks (Chapter 5); Mr P. G. G. Jackson (Chapter 6); several colleagues in the Farm Buildings Departments of the Aberdeen School of Agriculture (Chapter 7); Dr A. M. Petchey (Chapter 10); Dr J. F. O'Grady (Chapter 11) and Dr V. R. Fowler (Chapter 12).

It should be made clear that these gentlemen merely commented on the views of the authors, and while many of their constructive criticisms have been incorporated, the authors accept full responsibility for the statements and recommendations made.

We acknowledge gratefully the interchange of ideas with many pigkeepers and their stockmen, with many colleagues in the agricultural advisory and veterinary fields and with students. These interactions have assisted greatly in evaluating critically existing systems and in evolving and crystallising a basis for improved practices in many aspects of weaner production. We would like to make particular mention of Mr Arthur Simmers, and the managers and farm staff of J. A. Simmers & Sons, for opportunities afforded in the past few years for testing many of our ideas on their large commercial units.

Our thanks are due to Professor J. B. Owen, Head of the Animal Production and Health Group of the Aberdeen School of Agriculture, for guidance and encouragement throughout the preparation of this book.

We are grateful to Professor/Principal J. R. Raeburn of the Aberdeen School of Agriculture and, through him, to the School, its farms and supervisory and technical staff for the opportunities for research and investigation, many of the findings of which are reported here. Particular thanks are due to Mr George MacConnachie who provided most valued and skilled assistance over many years in collecting much of the data contained in Chapters 6, 8 and 9.

Finally, our thanks are due to Miss H. Jupp and her colleagues and Mrs W. G. Law for most valued typing services, to the late Mr Janis Stauvers, Miss Morag Young, Mr Ian Robb, Mr John Wiseman, Miss Fiona MacNeil, Mr Jim Reaper and Mr W. T. Findlater for the graphics and photographs. And to Mr Mervyn Dias and Mr Andrew Faulks for valued assistance in collating and checking the proofs and diagrams.

Acknowledgment is made under the tables and figures of the source except where the information presented is based on our own studies.

INDEX

Lisa Edwards